Transactional Memory

2nd edition

Transactional Memory, 2nd edition

Tim Harris, James Larus, and Ravi Rajwar

ISBN: 978-3-031-00600-5   paperback
ISBN: 978-3-031-01728-5   ebook

DOI 10.1007/987-3-031-01728-2

A Publication in the Springer series
SYNTHESIS LECTURES ON COMPUTER ARCHITECTURE

Lecture #11
Series Editor: Mark D. Hill, *University of Wisconsin*
Series ISSN
Synthesis Lectures on Computer Architecture
Print 1935-3235   Electronic 1935-3243

Synthesis Lectures on Computer Architecture

Editor
Mark D. Hill, *University of Wisconsin*

Synthesis Lectures on Computer Architecture publishes 50- to 100-page publications on topics pertaining to the science and art of designing, analyzing, selecting and interconnecting hardwarecomponents to create computers that meet functional, performance and cost goals. The scope will largely follow the purview of premier computer architecture conferences, such as ISCA, HPCA, MICRO, and ASPLOS.

Transactional Memory, 2nd edition
Tim Harris, James Larus, and Ravi Rajwar
2010

Computer Architecture Performance Evaluation Models
Lieven Eeckhout
2010

Introduction to Reconfigured Supercomputing
Marco Lanzagorta, Stephen Bique, and Robert Rosenberg
2009

On-Chip Networks
Natalie Enright Jerger and Li-Shiuan Peh
2009

The Memory System: You Can't Avoid It, You Can't Ignore It, You Can't Fake It
Bruce Jacob
2009

Fault Tolerant Computer Architecture
Daniel J. Sorin
2009

The Datacenter as a Computer: An Introduction to the Design of Warehouse-Scale Machines
Luiz André Barroso and Urs Hölzle
2009

Transactional Memory

2nd edition

Tim Harris
Microsoft Research

James Larus
Microsoft Research

Ravi Rajwar
Intel Corporation

SYNTHESIS LECTURES ON COMPUTER ARCHITECTURE #11

ABSTRACT

The advent of multicore processors has renewed interest in the idea of incorporating transactions into the programming model used to write parallel programs. This approach, known as transactional memory, offers an alternative, and hopefully better, way to coordinate concurrent threads. The ACI (atomicity, consistency, isolation) properties of transactions provide a foundation to ensure that concurrent reads and writes of shared data do not produce inconsistent or incorrect results. At a higher level, a computation wrapped in a transaction executes atomically - either it completes successfully and commits its result in its entirety or it aborts. In addition, isolation ensures the transaction produces the same result as if no other transactions were executing concurrently. Although transactions are not a parallel programming panacea, they shift much of the burden of synchronizing and coordinating parallel computations from a programmer to a compiler, to a language runtime system, or to hardware. The challenge for the system implementers is to build an efficient transactional memory infrastructure. This book presents an overview of the state of the art in the design and implementation of transactional memory systems, as of early spring 2010.

KEYWORDS

transactional memory, parallel programming, concurrent programming, compilers, programming languages, computer architecture, computer hardware, nonblocking algorithms, lock-free data structures, cache coherence, synchronization

Contents

Preface

This book presents an overview of the state of the art in transactional memory, as of early 2010. Substantial sections of this book have been revised since the first edition. There has been a vast amount of research on TM in the last three years (quantitatively, 210 of the 351 papers referred to in this book were written in 2007 or later). This work has expanded the range of implementation techniques that have been explored, the maturity of many of the implementations, the experience that researchers have writing programs using TM, and the insights from formal analysis of TM algorithms and the programming abstractions built over them.

At a high level, readers familiar with the first edition will notice two broad changes:

First, we have expanded the discussion of programming with TM to form two chapters. This reflects a separation between the lower level properties of transactions (Chapter 2) versus higher-level language constructs (Chapter 3). In early work, these notions were often combined with research papers introducing both a new TM algorithm and a new way of exposing it to the programmer. There is now a clearer separation, with common TM algorithms being exposed to programmers through many different interfaces, and with individual language features being implemented over different TMs.

The second main difference is that we have re-structured the discussions of STM (Chapter 4) and HTM (Chapter 5) so that they group work thematically rather than considering work chronologically on a paper-by-paper basis. In each case, we focus on detailed case studies that we feel are representative of major classes of algorithms or of the state-of-the-art. We try to be complete, so please let us know if there is work that we have omitted.

This book does not contain the answers to many questions. At this point in the evolution of the field, we do not have enough experience building and using transactional memory systems to prefer one approach definitively over another. Instead, our goal in writing this book is to raise the questions and provide an overview of the answers that others have proposed. We hope that this background will help consolidate and advance research in this area and accelerate the search for answers.

In addition, this book is written from a practical viewpoint, with an emphasis on the design and implementation of TM systems, and their integration into programming languages. Some of the techniques that we describe come from research that was originally presented in a more formal style; we provide references to the original papers, but we do not attempt a formal presentation in this book. A forthcoming book examines TM from a theoretical viewpoint [117].

There is a large body of research on techniques like thread-level speculation (TLS) and a history of cross-fertilization between these areas. For instance, Ding *et al.*'s work on value-based validation inspired techniques used in STM systems [88], whereas STM techniques using eager version management inspired Oancea *et al.*'s work on in-place speculation [234]. Inevitably, it is difficult to delineate exactly what work should be considered "TM" and what should not. Broadly speaking, we focus on work providing shared-memory synchronization between multiple explicit threads; we try, briefly, to identify links with other relevant work where possible.

The bibliography that we use is available online at `http://www.cs.wisc.edu/trans-memory/biblio/index.html`; we thank Jayaram Bobba and Mark Hill for their help in maintaining it, and we welcome additions and corrections.

Tim Harris, James Larus, and Ravi Rajwar
June 2010

Acknowledgments

This book has benefited greatly from the assistance of a large number of people who discussed transactional memory in its many forms with the authors and influenced this book—both the first edition and this revised edition. Some people were even brave enough to read drafts and point out shortcomings (of course, the remaining mistakes are the authors' responsibility).

Many thanks to: Adam Welc, Al Aho, Ala Alameldeen, Amitabha Roy, Andy Glew, Annette Bieniusa, Arch Robison, Bryant Bigbee, Burton Smith, Chris Rossbach, Christos Kotselidis, Christos Kozyrakis, Craig Zilles, Dan Grossman, Daniel Nussbaum, David Callahan, David Christie, David Detlefs, David Wood, Ferad Zyulkyarov, Gil Neiger, Goetz Graefe, Haitham Akkary, James Cownie, Jan Gray, Jesse Barnes, Jim Rose, João Cachopo, João Lourenço, Joe Duffy, Justin Gottschlich, Kevin Moore, Konrad Lai, Kourosh Gharachorloo, Krste Asanovic, Mark Hill, Mark Moir, Mark Tuttle, Martín Abadi, Maurice Herlihy, Michael Scott, Michael Spear, Milind Girkar, Milo Martin, Nathan Bronson, Nir Shavit, Pascal Felber, Paul Petersen, Phil Bernstein, Richard Greco, Rob Ennals, Robert Geva, Sanjeev Kumar, Satnam Singh, Scott Ananian, Shaz Qadeer, Simon Peyton Jones, Steven Hand, Suresh Jagannathan, Suresh Srinivas, Tony Hosking, Torvald Riegel, Vijay Menon, Vinod Grover, and Virendra Marathe.

Tim Harris, James Larus, and Ravi Rajwar
June 2010

CHAPTER 1

Introduction

1.1 MOTIVATION

As Bruce Springsteen sings, "good times got a way of coming to an end". Lost in the clamor of the Y2K nonevent and the .com boom and bust, a less heralded but more significant milestone occurred. Around 2004, 50 years of exponential improvement in the performance of sequential computers ended [237]. Although the quantity of transistors on a chip continues to follow Moore's law (doubling roughly every two years), it has become increasingly difficult to continue to improve the performance of sequential processors. Simply raising the clock frequency to increase performance is difficult due to power and cooling concerns. In the terminology of Intel's founder, Andrew Grove, this is an inflection point—a "time in the life of a business when its fundamentals are about to change" [116].

Industry's response to stalled sequential performance was to introduce single-chip, parallel computers, variously known as "chip multiprocessors", "multicore", or "manycore" systems. The architecture of these computers puts two or more independent processors on a single chip and connects them through a shared memory. The architecture is similar to shared-memory multiprocessors.

This parallel architecture offers a potential solution to the problem of stalled performance growth. The number of processors that can be fabricated on a chip will continue to increase at the Moore's law rate, at least for the next few generations. As the number of processors on a chip doubles, so does the peak number of instructions executed per second—without increasing clock speed. This means that the performance of a well-formulated parallel program will also continue to improve at roughly Moore's law rate. Continued performance improvement permits a program's developers to increase its functionality by incorporating sophisticated, new features—the dynamic that has driven the software industry for a long time.

1.1.1 DIFFICULTY OF PARALLEL PROGRAMMING

Unfortunately, despite more than 40 years' experience with parallel computers, programming them has proven to be far more difficult than sequential programming. Parallel algorithms are more difficult to formulate and prove correct than sequential algorithms. A parallel program is far more difficult to design, write, and debug than an equivalent sequential program. The non-deterministic bugs that occur in parallel programs are notoriously difficult to find and remedy. Finally, to add insult to injury, parallel programs often perform poorly. Part of these difficulties may be attributable to the exotic nature of parallel programming, which was of interest to only a small community, was not widely investigated or taught by academics, and was ignored by most software vendors.

However, this explanation only addresses part of the problem with parallel programming; it is fundamentally more difficult than sequential programming, and people have a great deal of difficulty keeping track of multiple events occurring at the same time. Psychologists call this phenomena "attention" and have been studying it for a century. A seminal experiment was Cherry's dichotic listening task, in which a person was asked to repeat a message heard in one ear, while ignoring a different message played to the other ear [60]. People are very good at filtering the competing message because they attend to a single channel at a time.

Parallelism and nondeterminacy greatly increase the number of items that a software developer must keep in mind. Consequently, few people are able to systematically reason about a parallel program's behavior. Consider an example. Professor Herlihy of Brown University has observed that implementing a queue data structure is a simple programming assignment in an introductory programming course. However, the design of queues that allow concurrent operations on both ends remains an active research topic, with designs tailored to different APIs (e.g., whether or not push and pop need to be supported on both ends of the queue), or taking different approaches in boundary conditions (e.g., whether or not items can appear duplicated or missed) [22; 223; 224].

The design of a parallel analogue, which allows concurrent threads to enqueue and dequeue elements, is a publishable result because of the difficulty of coordinating concurrent access and handling the boundary conditions [223]. In addition, program analysis tools, which compensate for human failings by systematically identifying program defects, find parallel code to be provably more difficult to analyze than sequential code [257].

Finally – and a primary motivation for the strong interest in transactional memory – the programming models, languages, and tools available to a parallel programmer have lagged far behind those for sequential programs. Consider two prevalent parallel programming models: data parallelism and task parallelism.

Data parallelism is an effective programming model that applies an operation simultaneously to an aggregate of individual items [153]. It is particularly appropriate for numeric computations, which use matrices as their primary data structures. Programs often manipulate a matrix as an aggregate, for example, by adding it to another matrix. Scientific programming languages, such as High Performance Fortran (HPF) [201], directly support data parallel programming with a collection of operators on matrices and ways to combine these operations. Parallelism is implicit and abundant in data parallel programs. A compiler can exploit the inherent concurrency of applying an operation to the elements of an aggregate by partitioning the work among the available processors. This approach shifts the burden of synchronization and load balancing from a programmer to a compiler and runtime system. Unfortunately, data parallelism is not a universal programming model. It is natural and convenient in some settings [153] but difficult to apply to most data structures and programming problems.

The other common programming model is *task parallelism*, which executes computations on separate threads that are coordinated with explicit synchronization such as fork-join operations, locks, semaphores, queues, etc. This unstructured programming model imposes no restrictions on the code that each thread executes, when or how threads communicate, or how tasks are assigned to threads. The model is a general one, capable of expressing all forms of parallel computation. It, however, is very difficult to program correctly. In many ways, the model is at the same (low) level of abstraction as the underlying computer's hardware; in fact, processors directly implement many of the constructs used to write this type of program.

1.1.2 PARALLEL PROGRAMMING ABSTRACTIONS

A key shortcoming of task parallelism is its lack of effective mechanisms for abstraction and composition—computer science's two fundamental tools for managing complexity. An *abstraction* is a simplified view of an entity, which captures the features that are essential to understand and manipulate it for a particular purpose. People use abstraction all the time. For example, consider an observer "Jim" and a dog "Sally" barking from the backyard across the street. Sally is Jim's abstraction of the dog interrupting his writing of this book. In considering her barking, Jim need not remember that Sally is actually a one-year-old Golden Retriever and, certainly, Jim does not think of her as a quadruped mammal. The latter specifics are true, but not germane to Jim's irritation at the barking. Abstraction hides irrelevant detail and complexity, and it allows humans (and computers) to focus on the aspects of a problem relevant to a specific task.

Composition is the ability to put together two entities to form a larger, more complex entity, which, in turn, is abstracted into a single, composite entity. Composition and abstraction are closely related since details of the underlying entities can be suppressed when manipulating the composite product. Composition is also a common human activity. Consider an engineered artifact such as a car, constructed from components such as an engine, brakes, body, etc. For most purposes, the abstraction of a car subsumes these components and allows us to think about a car without considering the details explored in automobile enthusiast magazines.

Modern programming languages support powerful abstraction mechanisms, as well as rich libraries of abstractions for sequential programming. Procedures offer a way to encapsulate and name a sequence of operations. Abstract datatypes and objects offer a way to encapsulate and name data structures as well. Libraries, frameworks, and design patterns collect and organize reusable abstractions that are the building blocks of software. Stepping up a level of abstraction, complex software systems, such as operating systems, databases or middleware, provide the powerful, generally useful abstractions, such as virtual memory, file systems, or relational databases used by most software. These abstraction mechanisms and abstractions are fundamental to modern software development which increasingly builds and reuses software components, rather than writing them from scratch.

Parallel programming lacks comparable abstraction mechanisms. Low-level parallel programming models, such as threads and explicit synchronization, are unsuitable for constructing abstractions because explicit synchronization is not composable. A program that uses an abstraction containing explicit synchronization must be aware of its details, to avoid causing races or deadlocks.

Here is an example. Consider a hashtable that supports thread-safe `Insert` and `Remove` operations. In a sequential program, each of these operations can be an abstraction. One can fully specify their behavior without reference to the hashtable's implementation. Now, suppose that in a parallel program, we want to construct a new operation, call it `Move`, which deletes an item from one hashtable and inserts it into another table. The intermediate state, in which neither table contains the item, must not be visible to other threads. Unless this requirement influences the implementation, there is no way to compose `Insert` and `Remove` operations to satisfy this requirement since they lock the table only for the duration of the individual operations. Fixing this problem requires new methods such as `LockTable` and `UnlockTable`, which break the hashtable abstraction by exposing an implementation detail. Moreover, these methods are error prone. A client that locks more than one table must be careful to lock them in a globally consistent order (and to unlock them!), to prevent deadlock.

The same phenomenon holds for other forms of composition. Suppose that a procedure `p1` waits for one of two input queues to produce data, making use of a library function `WaitAny` that takes a list of queues to wait for. A second procedure `p2` might do the same thing on two different queues. We cannot apply `WaitAny` to `p1` and `p2` to wait on any of the four queues: the inputs to `WaitAny` must be queues, but `p1` and `p2` are procedures. This is a fundamental loss of compositionality. Instead, programmers use awkward programming techniques, such as collecting queues used in lower-level abstractions, performing a single top-level `WaitAny`, and then dispatching back to an appropriate handler. Again, two individually correct abstractions, `p1` and `p2`, cannot be composed into a larger one; instead, they must be ripped apart and awkwardly merged, in direct conflict with the goals of abstraction.

1.2   DATABASE SYSTEMS AND TRANSACTIONS

While parallelism has been a difficult problem for general-purpose programming, database systems have successfully exploited parallel hardware for decades. Databases achieve good performance by executing many queries simultaneously and by running queries on multiple processors when possible. Moreover, the database programming model ensures that the author of an individual query need not worry about this parallelism. Many have wondered if the programming model used by databases, with its relative simplicity and widespread success, could also function as a more general, parallel programming model.

At the heart of the programming model for databases is a *transaction*. A transaction specifies a program semantics in which a computation executes as if it was the only computation accessing the database. Other computations may execute simultaneously, but the model restricts the allowable interactions among the transactions, so each produces results indistinguishable from the situation in which the transactions run one after the other. This model is known as *serializability*. As a consequence, a programmer who writes code for a transaction lives in the simpler, more familiar sequential programming world and only needs to reason about computations that start with the final results of other transactions. Transactions allow concurrent operations to access a common database and still produce predictable, reproducible results.

Transactions are implemented by an underlying database system or transaction processing monitor, both of which hide complex implementations behind a relatively simple interface [31; 113; 256]. These systems contain many sophisticated algorithms, but a programmer only sees a simple programming model that subsumes most aspects of concurrency and failure. Moreover, the abstract specification of a transaction's behavior provides a great deal of implementation freedom and allows the construction of efficient database systems.

Transactions offer a proven abstraction mechanism in database systems for constructing reusable parallel computations. A computation executed in a transaction need not expose the data it accesses or the order in which these accesses occur. Composing transactions can be as simple as executing subtransactions in the scope of a surrounding transaction. Moreover, coordination mechanisms provide concise ways to constrain and order the execution of concurrent transactions.

The advent of multicore processors has renewed interest in an old idea, of incorporating transactions into the programming model used to write parallel programs—building on ideas from languages such as Argus that have provided transactions to help structure distributed algorithms [196]. While programming-language transactions bear some similarity to these transactions, the implementation and execution environments differ greatly, as operations in distributed systems typically involves network communication, and operations to a transactional databases typically involve disk accesses. In contrast, programs typically store data in memory. This difference has given this new abstraction its name, *transactional memory* (TM).

1.2.1   WHAT IS A TRANSACTION?

A transaction is a sequence of actions that appears indivisible and instantaneous to an outside observer. A database transaction has four specific attributes: failure atomicity, consistency, isolation, and durability—collectively known as the ACID properties.

Atomicity requires that all constituent actions in a transaction complete successfully, or that none of these actions appear to start executing. It is not acceptable for a constituent action to fail and for the transaction to finish successfully. Nor is it acceptable for a failed action to leave behind evidence that it executed. A transaction that completes successfully *commits* and one that fails *aborts*. In this book, we will call this property *failure atomicity*, to distinguish it from a more expansive notion of *atomic execution*, which encompasses elements of other ACID properties.

The next property of a transaction is *consistency*. The meaning of consistency is entirely application dependent, and it typically consists of a collection of invariants on data structures. For example, an invariant might require that a `numCustomers` value contains the number of items in the `Customer` table, or that the `Customer` table does not contain duplicate entries.

If a transaction modifies the state of the world, then its changes should start from one consistent state and leave the database in another consistent state. Later transactions may have no knowledge of which transactions executed earlier, so it is unrealistic to expect them to execute properly if the invariants that they expect are not satisfied. Maintaining consistency is trivially satisfied if a transaction aborts, since it then does not perturb the consistent state that it started in.

The next property, called *isolation*, requires that transactions do not interfere with each other while they are running—regardless of whether or not they are executing in parallel. We will explore the semantics of transactions in the next chapter. This property obviously makes transactions an attractive programming model for parallel computers.

The final property is *durability*, which requires that once a transaction commits, its result be permanent (i.e., stored on a durable media such as disk) and available to all subsequent transactions.

1.3   TRANSACTIONAL MEMORY

In 1977, Lomet observed that an abstraction similar to a database transaction might make a good programming language mechanism to ensure the consistency of data shared among several processes [199]. The paper did not describe a practical implementation competitive with explicit synchronization, and so the idea lay fallow until 1993 when Herlihy and Moss [148] proposed hardware-supported transactional memory, and Stone *et al.* [309] proposed an atomic multi-word operation known as "Oklahoma Update" (a reference to the song "All er Nothin'" from the Rodgers and Hammerstein musical *Oklahoma!*). In recent years, there has been a huge ground swell of interest in both hardware and software systems for implementing transactional memory.

The basic idea is very simple. The properties of transactions provide a convenient abstraction for coordinating concurrent reads and writes of shared data in a concurrent or parallel system. Today, this coordination is the responsibility of a programmer, who has only low-level mechanisms, such as locks, semaphores, mutexes, etc., to prevent two concurrent threads from interfering. Even modern languages such as Java and C# provide only a slightly higher level construct, a monitor, to prevent concurrent access to an object's internal data. As discussed previously, these low-level mechanisms are difficult to use correctly and are not composable.

Transactions provide an alternative approach to coordinating concurrent threads. A program can wrap a computation in a transaction. Failure atomicity ensures the computation completes successfully and commits its result in its entirety or aborts. In addition, isolation ensures that the transaction produces the same result as it would if no other transactions were executing concurrently.

Although isolation appears to be the primary guarantee of transactional memory, the other properties, failure atomicity and consistency, are important. If a programmer's goal is a correct program, then consistency is important since transactions may execute in unpredictable orders. It would be difficult to write correct code without the assumption that a transaction starts executing in a consistent state. Failure atomicity is a key part of ensuring consistency. If a transaction fails, it could leave data in an unpredictable and inconsistent state that would cause subsequent transactions to fail. Moreover, a mechanism used to implement failure atomicity, reverting data to an earlier state, turns out to be very important for implementing certain types of concurrency control.

In this section, we provide a brief overview of the main topics that we study in the remainder of the book; we sketch a basic low-level programming interface for TM (Section 1.3.1) and the way in which it can be exposed to programmers in the more palatable form of `atomic` blocks (Section 1.3.2). We sketch a software implementation (Section 1.3.3) and discuss current performance results. We

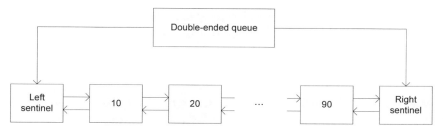

Figure 1.1: A double-ended queue, implemented as a doubly-linked list of elements strung between two sentinel nodes.

then discuss how TM can be supported by hardware (Section 1.3.4). We return to each of these topics as complete chapters.

Finally, we discuss some overall questions about exactly what kinds of workload are suited to TM (Section 1.3.5), how TM relates to database transactions (Section 1.3.6), and we provide pointers to the current TM systems that are available (Section 1.3.7).

1.3.1   BASIC TRANSACTIONAL MEMORY

Chapter 2 introduces TM, sketches its use, and presents a broad taxonomy of design choices for software and hardware TM systems. We refer to this taxonomy throughout the book when introducing different TM implementations.

As an running example, we return to the challenge from Professor Herlihy of building a scalable double-ended queue using a doubly-linked list (Figure 1.1). Concrete TM interfaces differ between implementations, but as an initial illustration, let us consider how the PushLeft operation could be implemented using a stylized TM system:

```
void PushLeft(DQueue *q, int val) {
  QNode *qn = malloc(sizeof(QNode));
  qn->val = val;
  do {
    StartTx();
    QNode *leftSentinel = ReadTx(&(q->left));
    QNode *oldLeftNode = ReadTx(&(leftSentinel->right));
    WriteTx(&(qn->left), leftSentinel);
    WriteTx(&(qn->right), oldLeftNode);
    WriteTx(&(leftSentinel->right), qn);
    WriteTx(&(oldLeftNode->left), qn);
  } while (!CommitTx());
}
```

This code fragment attempts to allocate a new QNode object and to splice it in to a doubly-linked-list representing the queue itself. The structure of the code broadly follows the equivalent sequential program: it allocates the `QNode` and then makes updates to the various fields of the new object and to the links from the existing queue. The memory accesses themselves are performed by `ReadTx` and `WriteTx` operations that are provided by the TM implementation, and this complete series of accesses is bracketed by a `StartTx()`...`CommitTx()` pair that delimits the scope of the transaction. We return to the details of this TM interface in Chapter 2.

This is a fundamentally different level of abstraction from trying to write the same operation using manual locking: the programmer has not needed to indicate where to acquire or release locks, or to identify which operations `PushLeft` may be allowed to execute concurrently with. Both of these are the responsibility of the TM implementation.

There are two main mechanisms that the TM implementation needs to provide. First, it needs to manage the tentative work that a transaction does while it executes. For instance, in this case, the TM needs to track the updates that are being made to the `left` and `right` fields of the `qn` object and to the existing fields of the nodes in the queue. Typically, this is done by the TM system either (*i*) writing directly to memory while maintaining information about the values that it overwrites ("eager versioning" because the TM makes its writes as soon as possible) or (*ii*) building up a private buffer holding the updates that a transaction wishes to make before writing them out to memory if the transaction commits successfully ("lazy versioning", because the updates are only made at the end, and only if the commit is successful).

The second mechanism that the TM must provide is a way to ensure isolation between transactions; the TM needs to detect conflicts that occur and to resolve these conflicts so that concurrent transactions appear to execute one-after-the-other rather than leaving a jumbled mess in memory. A TM system using "eager conflict detection" identifies possible conflicts while transactions are running. For instance, if two threads attempt to call `PushLeft` on the same queue at the same time, then a conflict occurs when they both try to update the `leftSentinel` object to refer to their new nodes: one of the transactions needs to be aborted, and the other can be allowed to continue. An alternative approach is "lazy conflict detection" in which transactions continue to run speculatively, and conflicts are detected only when they try to commit: in this case one of the `CommitTx` operations would return `false` because the other transaction has successfully committed.

The trade offs between these different approaches are a recurring topic throughout the book and, indeed, throughout research on TM implementations.

1.3.2   BUILDING ON BASIC TRANSACTIONS

Chapter 3 discusses how TM can be integrated into a high-level programming language. A frequent approach is to provide `atomic` blocks. For instance, the example `PushLeft` operation might be written:

```
void PushLeft(DQueue *q, int val) {
  QNode *qn = malloc(sizeof(QNode));
  qn->val = val;
  atomic {
    QNode *leftSentinel = q->left;
    QNode *oldLeftNode = leftSentinel->right;
    qn->left = leftSentinel;
    qn->right = oldLeftNode;
    leftSentinel->right = qn;
    oldLeftNode->left = qn;
  }
}
```

This approach eliminates a lot of the boilerplate associated with using TM directly: the programmer identifies the sections of code that should execute atomically, and the language implementation introduces operations such as ReadTx and WriteTx where necessary.

Furthermore, although they are not a focus of this current book, implementations of atomic blocks have been developed that rely solely on static analyses to infer sets of locks that the program should acquire [59; 74; 129; 151; 214]: atomic blocks provide a high-level abstraction that can implemented in different ways on different systems.

Chapter 3 also discusses some of the key challenges in integrating TM into existing software tool-chains and, in particular, it looks at the question of how TM can coexist with other programming abstractions. For example, what happens if a program attempts to write a message to the screen within the middle of an atomic block? One might expect that the message should just be printed once, but a naïve transactional implementation might write the output multiple times if the atomic block's implementation needs to be attempted more than once before it commits. We discuss different options for defining what the correct behavior should be in cases like this.

1.3.3   SOFTWARE TRANSACTIONAL MEMORY

Chapter 4 describes software transactional memory (STM) implementation techniques, focusing on approaches that can be implemented on current mainstream processors. This has been a rich vein of research over the last decade, and numerous alternatives have been explored.

Returning to the example PushLeft function, one class of STM systems combines automatic locking of locations that are updated (to avoid conflicting updates by concurrent writing transactions) along with the use of per-object version numbers (to detect conflicts between a reader and any concurrent writers).

For example, if one thread executes the PushLeft operation in isolation then its transaction would start by recording the version number in the q object and then recording the version number in the leftSentinel object. These version numbers would be stored in a *read-log* that the transaction maintains in thread-private storage. The transaction then attempts to write to the qn object, and so it would acquire an exclusive lock on this object and add it to its *write-log*. As the transaction makes updates, it would use eager version management (writing the updates directly to the objects themselves and recording the values that it overwrites into a third thread-private *undo-log*). Finally, when the transaction tries to commit, it would check the version numbers for each entry in its read-log to make sure that there have not been any conflicting updates. Since there have been none, it would increment the version numbers on the objects in its write-log, before releasing the locks on these objects.

The logs provide the STM system with all of the information necessary to detect conflicts and to resolve them. The version numbers in the read-log allow a thread to detect whether a concurrent thread has updated an object that it has read from (because an update will have incremented the object's version number). The values recorded in the undo log allow a thread to undo a transaction if this kind of conflict occurs.

As we discuss in Chapter 4, there is a very large space of different possible STM designs, and this simple version-number-based approach is by no means the state of the art.

When comparing alternative STM systems, it is useful to distinguish different properties that they might have. Some STM systems aim for *low sequential overhead* (in which code running inside a transaction is as fast as possible), others aim for *good scalability* (in which a parallel workload using transactions can improve in performance as processors are added) or for *strong progress guarantees* (e.g., to provide nonblocking behavior). STM systems also differ substantially in terms of the programming semantics that they offer—e.g., the extent to which memory locations may be accessed transactionally at some times and non-transactionally at other times.

In addition, when evaluating TM systems, we must consider that software implementations on today's multiprocessor systems execute with much higher interprocessor communication latencies, which may favor computationally expensive approaches that incur less synchronization or cache traffic. Future systems may favor other trade offs. For instance, in the context of reader-writer locks, Dice and Shavit have shown how CMPs favor quite different design choices from traditional multiprocessors [86]. Differences in the underlying computer hardware can greatly affect the performance of an STM system—e.g., the memory consistency model that the hardware provides and the cost of synchronization operations when compared with ordinary memory accesses.

The extent to which STM systems can be fast enough for use in practice remains a contentious research question in itself—after all, in the simple system we described above, the STM system introduces additional work maintaining logs and performing synchronization operations on objects. The empirical results from Cascaval *et al.* [52] and Dragojević *et al.* [92] provide two recent sets of observations that draw different conclusions.

For these reasons, in this survey book, we generally omit detailed discussion of the performance of specific TM algorithms. However, to provide some initial quantitative examples, we briefly review the results from Dragojević *et al.*'s study. Their study examines the number of threads that are required for an application using STM to out-perform a sequential version of the same program; this defines a break-even point beyond which the parallel program has the potential to be faster than the sequential program [92]. Dragojević *et al.* report that, on a SPARC processor, 8/17 workloads reached their break-even point with 2 threads, 14/17 workloads had reached their break-even point by 4 threads, and 16/17 by 8 threads.

1.3.4   HARDWARE TRANSACTIONAL MEMORY

Whether or not STM is fast enough in itself, there is clearly scope for significant performance improvements through hardware support. Chapter 5 describes Hardware Transactional Memory (HTM) implementation techniques—including systems that provide complete implementations of TM in hardware, systems that allow HW transactions to coexist with SW transactions, and systems that provide hardware extensions to speed up parts of an STM implementation.

Early HTM systems kept a transaction's modified state in a cache and used the cache coherence protocol to detect conflicts with other transactions. Recent HTM systems have explored using a processor's write buffer to hold transactional updates, or spilling transactional data into lower levels of the memory hierarchy or into software-managed memory.

We examine HTM systems at two different levels. First, there is the software programming model that they support: Do they require specific new instructions to be used when making a transactional memory access (akin to ReadTx and WriteTx) or, if a transaction is active, are all memory accesses implicitly transactional? Does the HTM require transaction boundaries to be identified explicitly, or does it infer them (e.g., based on ordinary lock-acquire operations)? Does the HTM automatically re-execute a transaction that experiences contention, or does it branch to a software handler that can perform an alternative operation? Importantly, does the HTM ensure that certain kinds of transactions are guaranteed to be able to commit in the absence of contention (say, those that access at most 4 memory locations?) All of these questions have important consequences for the programming abstractions that are built over HTM.

The second level we consider comprises the microarchitecture mechanisms that are used to support transactions—e.g., extensions to the cache coherence protocols for conflict detection or the development of entirely new memory systems that are based around transactional execution.

HTM systems typically provide primitive mechanisms that underlie the user-visible languages, compilers, and runtime systems. Software bridges the gap between programmers and hardware, which makes much of the discussion of STM systems, languages, and compilers relevant to HTM systems as well.

1.3.5   WHAT IS TRANSACTIONAL MEMORY GOOD FOR?

Much of the literature on TM systems focuses on implementation mechanisms and semantics, but leaves implicit the question of exactly where TM is an appropriate programming abstraction.

One recurring use for TM is in managing shared-memory data structures in which scalability is difficult to achieve via lock-based synchronization. The PushLeft operation we sketched for a double-ended queue is one example of this kind of use. It seems clear that even a modest form of TM is useful here—even a SW implementation with an explicit interface (like the ReadTx call in our PushLeft example), or a HW implementation that might limit transactions to only 2, 3, or 4 words. In this kind of example, the data structure implementation would be expected to be written by an expert programmer and most likely encapsulated in a standard library. Performance is key, and the ease of writing the library might be less important.

Another example where TM seems effective are graph algorithms in which the set of nodes that a thread accesses depends on the values that it encounters: with locks, a thread may need to be overly conservative (say, locking the complete graph before accessing a small portion of it), or a graph traversal may need to be structured with great care to avoid deadlocks if two threads might traverse the graph and need to lock nodes already in use by one another. With TM, a thread can access nodes freely, and the TM system is responsible for isolating separate transactions and avoiding deadlock.

Zyulkyarov *et al.* provide an illustration of a larger example from a TM-based implementation of the Quake game server [350]. When modeling the effect of a player's move in the game, the lock-based implementation needs to simulate the effect of the operation in order to determine which game objects it needs to lock. Having done that simulation, the game would lock the objects, check that the player's move is still valid, and then perform its effects. With TM, the structure of that code is simplified because the simulation step can be avoided.

In this example, transactions may need to be larger in size (say, dozens of memory accesses), and an interface like `ReadTx` may become cumbersome. For this kind of usage, there would be more importance placed on language support (e.g., `atomic` blocks) and more importance placed on portability of the program from one language implementation to another.

Transactions are not a panacea. In parallel software, the programmer must still divide work into pieces that can be executed on different processors. It is still (all too) easy to write an incorrect parallel program, even with transactional memory. For example, a programmer might write transactions that are too short—e.g., in the `Move` example, they might place `Insert` and `Remove` in two separate transactions, rather than in one combined transaction. Conversely, a programmer might write transactions that are too long—e.g., they might place two operations inside a transaction in one thread, when the intermediate state between the operations needs to be visible to another thread. A programmer might also simply use transactions incorrectly—e.g., starting a transaction but forgetting to commit it. Finally, and particularly with early software implementations, the performance of code executing within a transaction can be markedly slower than the performance of normal code. Since the purpose of using parallelism is typically to get a performance *increase*, the programmer must be sure that the performance gains of parallelism outweigh the penalties of synchronization.

1.3.6   DIFFERENCES BETWEEN DATABASE TRANSACTIONS AND TM

Transactions in memory differ from transactions in databases, and, consequently, they require new implementation techniques, a central topic of this book. The following differences are among the most important:

- Data in a traditional database resides on a disk, rather than in memory. Disk accesses take 5–10ms or, literally, time enough to execute millions of instructions. Databases can freely trade computation against disk access. Transactional memory accesses main memory, which incurs a cost of at most a few hundred instructions (and typically, in the case of a cache hit, only a handful of cycles). A transaction cannot perform much computation at a memory access. Hardware support is more attractive for TM than for database systems.

- Transactional memory is typically not durable in the sense that data in memory does not survive program termination. This simplifies the implementation of TM since the need to record data permanently on disk before a transaction commits considerably complicates a database system.

- A database provides the sole route of access to the data that it contains; consequently, the database implementer is free to choose how to represent the data, how to associate concurrency-control metadata with it, and so on. With TM, the programmer is typically able to perform normal memory accesses in addition to transactional ones—e.g., to access a piece of data directly from within one thread before starting to share it with other threads via transactions. Exactly which forms of such mixed-mode accesses are permitted is a recurring theme throughout this book.

- Transactional memory is a retrofit into a rich, complex world full of existing programming languages, paradigms, libraries, software, and operating systems. To be successful, transactional memory must coexist with existing infrastructure, even if a long-term goal may be to supplant portions of this world with transactions. Programmers will find it difficult to adopt transactional memory if it requires pervasive changes to programming languages, libraries, or operating systems—or if it compels a closed world, like databases, where the only way to access data is through a transaction.

1.3.7   CURRENT TRANSACTIONAL MEMORY SYSTEMS AND SIMULATORS

Numerous TM systems are now available. These are the ones that we are aware of (listed alphabetically):

- A simulator is available for ASF, a proposed AMD64 architecture extension for bounded-size transactions [61]. This is based on PTLSim, providing a detailed, cycle-accurate full-system simulation of a multi-core system. The simulator models the proposed hardware instructions. These are made available via C/C++ wrapper functions and macros. http://www.amd64.org/research/multi-and-manycore-systems.html

- CTL is a library-based STM implementation derived from an early version of the TL2 [83] algorithm. http://www-asc.di.fct.unl.pt/~jml/Research/Software

- Deuce STM[173] provides support for atomic methods in an unmodified implementation of the Java Virtual Machine. Methods are marked by an @Atomic attribute, and bytecode-to-bytecode rewriting is used to instrument them with STM operations. http://www.deucestm.org/

- DTMC, the Dresden TM Compiler, supports transactions in C/C++ based on a modified version of llvm-gcc and an additional LLVM compiler pass [61]. The system uses a C++ version of TinySTM, extended to include a range of different STM implementations. DTMC and TinySTM can also target the AMD ASF simulator. `http://tm.inf.tu-dresden.de`

- The IBM XL C/C++ for Transactional Memory compiler provides support for pragma-based atomic sections in C/C++ programs. It operates on the AIX operating system. `http://www.alphaworks.ibm.com/tech/xlcstm`. In addition, the source code of an STM implementation compatible with the compiler was released through the Amino Concurrency Building Blocks open source package. `http://amino-cbbs.sourceforge.net/`

- The Intel C++ STM compiler extends C++ with support for STM language extensions, including inheritance, virtual functions, templates, exception handling, failure atomicity, TM memory allocation, and irrevocable actions for legacy code & IO. A published ABI defines the interface between the compiler and the core TM implementation itself. `http://software.intel.com/en-us/articles/intel-c-stm-compiler-prototype-edition-20/`

- JVSTM is a Java library that implements a multi-versioned approach to STM that includes mechanisms for partial re-execution of failed transactions [45]. `http://web.ist.utl.pt/~joao.cachopo/jvstm/`

- Simulators are available for many variants of the LogTM [227] and LogTM-SE [40; 337] systems. These are released as part of the Wisconsin Multifacet GEMS simulation framework built on top of Virtutech Simics. They support different conflict detection and version management mechanisms, along with partial rollback, closed and open nesting. `http://www.cs.wisc.edu/gems/`. GEMS also supports the Adaptive Transactional Memory Test Platform (ATMTP) which models the Rock HTM instructions [82]. `http://www.cs.wisc.edu/gems/doc/gems-wiki/moin.cgi/ATMTP`

- MetaTM is a hardware transactional memory simulator. It operates as a module for the Virtutech Simics platform. The simulator can host TxLinux [271], which is a variant of the i386 Linux kernel designed to use MetaTM's hardware transactional memory model for its internal synchronization. `http://www.metatm.net`

- OSTM and WSTM are early nonblocking STM systems, released as part of the `lock-free-lib` package of lock-free data structures. OSTM and WSTM provide a library-based programming model, primarily aimed at data structure implementations. `http://www.cl.cam.ac.uk/research/srg/netos/lock-free/`

- RSTM is a comprehensive set of STM systems available from the Rochester Synchronization Group. The system is available as source code, and comprises a C++ package with 13 different STM library implementations, and a smart-pointer based API for relatively transparent access to STM without requiring compiler changes. The STM algorithms include the original

RSTM design [211], along with many recent designs such as variants of RingSTM [305], and NOrec [77]. RSTM supports numerous architectures (x86, SPARC, POWER, Itanium) and operating systems (Linux, Solaris, AIX, Win32, Mac). `http://code.google.com/p/rstm`

- STM.NET is an experimental extension of the .NET Framework to provide support for C# programmers to use `atomic` blocks. It is available from Microsoft in binary format, along with a set of example C# programs, and support for integration with Visual Studio 2008. `http://msdn.microsoft.com/en-us/devlabs/ee334183.aspx`

- The Sun C++ compiler with Transactional Memory supports a range of STM back-ends, including TL2 [83], SkySTM [188], HyTM [78] and PhTM [193]. The compiler is available in binary format, but the runtime system and additional TM implementations are available as source code by request from the Sun Labs Scalable Synchronization Research Group. `http://research.sun.com/scalable/`

- SwissTM [93] is a library-based STM system for C/C++ designed to support large transactions in addition to shorter-running ones used in some workloads. `http://lpd.epfl.ch/site/research/tmeval`

- TinySTM is a word-based STM implementation available from the University of Neuchatel. It is based on the LSA algorithm [262]. A Java LSA implementation is also available. `http://tmware.org`

- Implementations of TL2 [83] and subsequent algorithms are available for use with Tanger (an earlier version of DTMC). `http://mcg.cs.tau.ac.il/projects`

- Twilight STM extends library-based STMs for C and for Java with a notion of "twilight" execution during which transactions can attempt to detect and repair potential read inconsistencies and turn a failing transaction into a successful one. `http://proglang.informatik.uni-freiburg.de/projects/syncstm/`

- TxOS is a prototype version of Linux that extends the OS to allow composition of system calls into atomic, isolated operations [243]. TxOS supports transactional semantics for a range of resources, including the file system, pipes, signals, and process control. It runs on commodity hardware. `http://txos.code.csres.utexas.edu`

We aim for this list to be complete, and so we welcome suggestions for additions or amendments.

CHAPTER 2

Basic Transactions

This chapter presents transactional memory from the perspective of a low-level programmer or, equivalently, from the perspective of a compiler using TM to implement features in a high-level programming language. Most research on TM has focused on using it as a parallel programming construct, so this discussion will focus on that aspect as well, rather than using transactions for error recovery, real-time programming, or multitasking.

We focus on a simple TM interface, comprising operations for managing transactions and for performing memory accesses. Many extensions to this are important—for instance, to integrate transactions with existing programming libraries, or to allow transactions to express operations that need to block. We defer these extensions until Chapter 3. The terminology in this simplistic setting will provide a reference point when discussing particular implementations in detail in the case studies later in this book.

We use a stylized TM interface in which all transactional operations are explicit. This kind of interface can be supported both by HTM and STM—concretely, it is broadly similar to the original hardware design of Herlihy and Moss [148], or to the word-based STM of Fraser and Harris [106].

The stylized TM interface provides a set of operations for managing transactions:

```
// Transaction management
void StartTx();
bool CommitTx();
void AbortTx();
```

StartTx begins a new transaction in the current thread. CommitTx attempts to commit the current transaction; we say that it either *succeeds* and returns true or that it *fails*, aborting the transaction and returning false. In addition, many systems provide an AbortTx operation that *explicitly* aborts the current transaction, whether or not it has experienced a conflict.

The second set of operations is concerned with data access:

```
// Data access
T ReadTx(T *addr);
void WriteTx(T *addr, T v);
```

ReadTx takes the address addr of a value of type T and returns the transaction's view of the data at that address. WriteTx takes an address addr and a new value v, writing the value to the transaction's view of that address. In practice, different functions would be provided for different types T. Type parameters could be used in a language supporting templates or generic classes. We refer to the set of locations that a transaction has read from as its *read-set* and the set of locations that it has written to as its *write-set*.

Returning to the running example from Chapter 1, suppose that a program maintains a double-ended queue, represented by a doubly-linked list (Figure 1.1). The queue supports four operations:

```
void PushLeft(DQueue *q, int val);
int PopLeft(DQueue *q);
void PushRight(DQueue *q, int val);
int PopRight(DQueue *q);
```

Let us suppose that the queue can grow without bound, and that a pop operation can return a special value (say, −1) if it finds that the queue is empty. With TM, the implementation of the queue's operations can remain very close to a simple, sequential version of the queue. For instance, the sequential version of PushLeft might be:

```
void PushLeft(DQueue *q, int val) {
  QNode *qn = malloc(sizeof(QNode));
  qn->val = val;
  QNode *leftSentinel = q->left;
  QNode *oldLeftNode = leftSentinel->right;
  qn->left = leftSentinel;
  qn->right = oldLeftNode;
  leftSentinel->right = qn;
  oldLeftNode->left = qn;
}
```

In this example, the code creates a new QNode to hold the value, then it finds the queue's left-sentinel (leftSentinel), and the original left-most node in the queue (oldLeftNode). The transaction then initializes the fields of the QNode, and splices it between leftSentinel and oldLeftNode. As we showed in Chapter 1, the corresponding transactional code would be:

```
void PushLeft(DQueue *q, int val) {
  QNode *qn = malloc(sizeof(QNode));
  qn->val = val;
  do {
    StartTx();
    QNode *leftSentinel = ReadTx(&(q->left));
    QNode *oldLeftNode = ReadTx(&(leftSentinel->right));
    WriteTx(&(qn->left), leftSentinel);
    WriteTx(&(qn->right), oldLeftNode);
    WriteTx(&(leftSentinel->right), qn);
    WriteTx(&(oldLeftNode->left), qn);
  } while (!CommitTx());
}
```

The body of the operation has been wrapped in a StartTx...CommitTx pair, and the memory accesses themselves have been expanded to use ReadTx and WriteTx. This structure is typical with TM: the StartTx...CommitTx operations bracket a series of steps which update a data structure from one consistent state to another, and this is repeated until CommitTx succeeds.

Unlike a lock-based implementation of `PushLeft`, when using TM, it is not necessary to explicitly acquire and release locks; this is all handled by the TM implementation. In fact, a lock-based implementation of the algorithm would be extremely difficult to write in a way that is correct and scalable. Using a single lock to protect the entire queue would prevent operations proceeding concurrently on the two ends. To provide some degree of scalability, we might try to use separate locks for the left and right ends of the queue, so that operations on the two ends can proceed in parallel. However, care is needed when the queue is almost empty because we would have to avoid removing the final element from the queue from both ends at the same time—and we would need to avoid the possibility of deadlock if the two locks are needed together.

Using TM, the syntax of `PushLeft` is clearly verbose, and much can be done to improve it. As we show in the next chapter, better language constructs can remove the need to write out operations like `ReadTx` and `WriteTx`. However, language support also blurs the question of exactly which sequence of operations is invoked on the TM—particularly, when the implementation is coupled with an optimizing compiler that might re-order memory accesses. Therefore, initially, we focus on an explicit TM interface to avoid any ambiguity.

In the remainder of the current chapter, we examine three aspects of TM. First, in Section 2.1, we look at the broad characteristics of different TM systems—for instance, how the tentative updates made by transactions are managed and how conflicts between transactions are detected.

In Section 2.2, we look at the semantics of TM operations; how they relate to database transactions and how notions from database transactions be extended to cope with problems that are unique to TM.

Finally, in Section 2.3, we look at progress guarantees for TM or, more typically, a *lack* of progress that can occur in workloads that provoke frequent conflicts between transactions. Without care, a basic implementation of TM can fail to make any progress at all, even on short examples like `PushLeft`—in the worst case, implementations can discover that two transactions conflict, force both of them to abort, and then re-execute both of them. This cycle can repeat, resulting in livelock. Some TM systems guarantee that this kind of livelock cannot occur; others leave it to the programmer to avoid livelock.

Somewhat sloppily, throughout this book, we use the word *transaction* to refer to a complete series of `StartTx`...`CommitTx` attempts, as well as to each of these individual attempts until a commit succeeds. Many TM systems automatically detect conflicts and re-execute failed transactions (even before a transaction invokes `CommitTx`). Consequently, this distinction is not always present in implementations.

2.1   TM DESIGN CHOICES

Even for the simple TM interface above, many alternative implementations are possible. We introduce the main design choices in this section, looking at optimistic versus pessimistic concurrency control (Section 2.1.1), eager versus lazy version management (Section 2.1.2), and eager versus lazy conflict detection (Section 2.1.3).

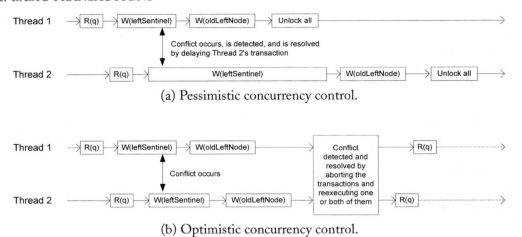

(a) Pessimistic concurrency control.

(b) Optimistic concurrency control.

Figure 2.1: Two approaches to concurrency control. R(x) indicates an attempt to read from x, and W(x) indicates an attempt to write to x.

2.1.1   CONCURRENCY CONTROL

A TM system requires synchronization to mediate concurrent accesses to data. We say that:

- A conflict *occurs* when two transactions perform conflicting operations on the same piece of data—either two concurrent writes, or a write from one transaction and a read from another.

- The conflict is *detected* when the underlying TM system determines that the conflict has occurred.

- The conflict is *resolved* when the underlying system or code in a transaction takes some action to avoid the conflict—e.g., by delaying or aborting one of the conflicting transactions.

These three events (conflict, detection, resolution) can occur at different times, but not in a different order—at least until systems predict or otherwise anticipate conflicts. Broadly, there are two approaches to concurrency control:

With *pessimistic concurrency control*, all three events occur at the same point in execution: when a transaction is about to access a location, the system detects a conflict, and resolves it. This type of concurrency control allows a transaction to claim exclusive ownership of data prior to proceeding, preventing other transactions from accessing it. For instance, with PushLeft, pessimistic concurrency control could be implemented by the TM locking the queue, locking the left sentinel, and locking the original left-most node before attempting to insert the new node (Figure 2.1(a)).

With *optimistic concurrency control* (Figure 2.1(b)), conflict detection and resolution can happen after a conflict occurs. This type of concurrency control allows multiple transactions to access data concurrently and to continue running even if they conflict, so long as the TM detects and resolves these conflicts before a transaction commits. This provides considerable implementation leeway—for instance, conflicts can be resolved by aborting a transaction or by delaying one of the conflicting transactions.

Both forms of concurrency control require care in their implementation to ensure that transactions make progress. Implementations of pessimistic concurrency control must avoid *deadlock*—e.g., if transaction TA holds a lock L1 and requests lock L2, while transaction TB holds L2 and requests L1. Deadlocks can be avoided by acquiring access in a fixed, predetermined order, or by using timeouts or other dynamic deadlock detection techniques to recover from them [176]. Conversely, some forms of optimistic concurrency control can lead to *livelock*—e.g., if transaction TA writes to location x then it may conflict with transaction TB and force TB to be aborted, whereupon TB may restart, write to x, forcing TA to be aborted. To avoid this problem, TM systems may need to use *contention management* to cause transactions to delay reexecution in the face of conflicts (Section 2.3.3), or they may need to ensure that a transaction is only aborted because of a conflict with a transaction that has actually committed.

If conflicts are frequent, then pessimistic concurrency control can be worthwhile: once a transaction has its locks, it is going to be able to run to completion. However, if conflicts are rare, optimistic concurrency control is often faster because it avoids the cost of locking and can increase concurrency between transactions. (In contrast, note that database systems often use pessimistic concurrency control even when conflicts are rare.)

Most of the TM systems we study in this book use optimistic concurrency control. However, hybrid approaches are also common, using pessimistic concurrency control between writes and optimistic concurrency control with reads (Section 4.2). Another common hybrid is to combine pessimistic synchronization of *irrevocable* transactions with an optimistic TM system [37; 300; 303; 308; 333]. An irrevocable transaction is one which the TM system guarantees will be able to run to completion without conflicts (Section 3.2). Irrevocability can be used to guarantee that a given transaction can finish—e.g., if it has experienced frequent contention in the past or performed an IO operation.

2.1.2 VERSION MANAGEMENT

TM systems require mechanisms to manage the tentative writes that concurrent transactions are doing ("version management"). The first general approach is *eager version management* [227]. This is also known as *direct update* because it means that the transaction directly modifies the data in memory. The transaction maintains an *undo-log* holding values that it has overwritten. This log allows the old values to be written back if the transaction subsequently aborts. Eager version management requires that pessimistic concurrency control be used for the transaction's writes; this is necessary because the transaction requires exclusive access to the locations if it is going to write to them directly.

The second general approach is *lazy version management*. This approach is also known as *deferred update* because the updates are delayed until a transaction commits. The transaction maintains its tentative writes in a transaction-private *redo-log*. A transaction's updates are buffered in this log, and a transaction's reads must consult the log so that earlier writes are seen. When a transaction commits, it updates the actual locations from these private copies. If a transaction aborts, it simply discards its redo-log.

2.1.3   CONFLICT DETECTION

The final main design choice in TM systems is how to detect conflicts. With pessimistic concurrency control, conflict detection is straightforward because a lock used in the implementation can only be acquired when it is not already held in a conflicting mode by another thread.

However, in TM systems using optimistic concurrency control, a very wide spectrum of techniques have been studied. In these systems, there is, typically, a *validation* operation with which the current transaction checks whether or not it has experienced conflicts; if validation succeeds, then the transaction's execution thus far could have occurred legitimately in some serial execution.

We can classify most approaches to conflict detection on three orthogonal dimensions:

The first dimension is the *granularity* of conflict detection: e.g., in HTM systems, conflicts may be detected at the level of complete cache lines, or, in STM system, conflicts may be detected at the level of complete objects. Most TM systems, therefore, involve some notion of *false conflicts* in which the implementation treats two transactions as conflicting even though they have accessed distinct locations—techniques using *value-based validation* provide one way to avoid false conflicts or to recover from false conflicts (Section 4.4.2).

The second dimension is the time at which conflict detection occurs:

- A conflict can be detected when a transaction declares its intent to access data (by "opening" or "acquiring" a location in many STMs) or at the transaction's first reference to the data. The term *eager conflict detection* is sometimes used to describe this approach [227].

- Conflicts can be detected on validation, at which point a transaction examines the collection of locations it previously read or updated, to see if another transaction has modified them. Validation can occur any time, or even multiple times, during a transaction's execution.

- A conflict can be detected on commit. When a transaction attempts to commit, it may (often must) validate the complete set of locations one final time to detect conflicts with other transactions. The term *lazy conflict detection* is sometimes used to describe this approach [227].

The final dimension is exactly which kinds of access are treated as conflicts:

- A system using *tentative* conflict detection identifies conflicts between concurrent transactions—e.g., if transaction TA has read from a location and transaction TB writes to the location, then this constitutes a conflict, even before either transaction commits.

- A system using *committed* conflict detection only considers conflicts between active transactions and those that have already committed—in this case, TA and TB can proceed in parallel, and a conflict only occurs when one or other of the transactions commits.

In practice, eager mechanisms are usually coupled with the detection of tentative conflicts between running transactions, while lazy mechanisms are typically coupled with detection of conflicts between a transaction that is trying to commit and any concurrent transactions that have already committed.

As with concurrency control, hybrid approaches are often used in a given TM system, and write–write and read–write conflicts may be treated differently. In addition, some TM systems provide abstractions for higher-level notions of conflict detection; we return to these high-level notions, and to the performance impact of different forms of conflict detection in Section 2.3.

2.2   SEMANTICS OF TRANSACTIONS

A programming abstraction with a simple, clean semantics helps programmers understand the programming construct, increases their chances of writing correct code, and facilitates detecting errors with programming tools. However, there is no agreed semantics of transactions although some attempts have been made [3; 4; 6; 89; 115; 119; 125; 126; 226; 282].

In this section, we describe the core approaches taken in the literature, concentrating initially on simple transactions and the link with databases (Section 2.2.1), then the form of consistency seen by a series of accesses within a single transaction (Section 2.2.2), the interactions between transactional and non-transactional memory accesses (Section 2.2.3–2.2.5) and different forms of nested transaction (Section 2.2.6).

Before discussing the semantics in detail, it is worthwhile distinguishing between two different levels of concurrency that are present in a system using TM. First, there can be concurrency between transactions: two transactions are concurrent unless one of them finishes before the other starts. Second, there can be concurrency between the individual TM operations that are used by the transactions: e.g., two different threads might execute ReadTx at the same time. Figure 2.2(a) illustrates this. In the figure, Thread 1's transaction is concurrent with Thread 2's transaction. In addition, the two StartTx operations run concurrently, and Thread 1's read from x is concurrent with Thread 2's read from u.

It is useful to decompose the problem of defining TM semantics by handling these two kinds of concurrency differently. In particular, we can focus on the core problem of defining the TM operations by requiring that they satisfy a property known as *linearizability* [150]. In order to be linearizable, the individual operations like StartTx and ReadTx must appear to take place atomically at some point between when the operation begins and when it finishes. Figure 2.2(b) illustrates this, mapping each of the TM operations onto a point on a time line at which they appear to run atomically.

Requiring linearizability lets us abstract over the low-level details of the TM system's internal concurrency control. However, we still need to consider what different sequential orderings of TM operations mean—e.g., exactly which values might be returned by Thread 2's read from u in Figure 2.2(b). This is known as a *sequential semantics* of TM [282].

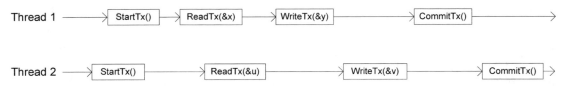

(a) Two levels of concurrency: the *transactions* being executed by the threads are concurrent, and some of the individual *TM operations* are also concurrent (e.g., the two calls to `StartTx`).

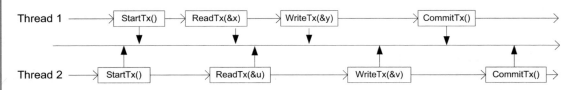

(b) Linearizability of TM operations: each operation appears to execute atomically (as shown by the solid arrows), at some point between its start and its completion.

Figure 2.2: Concurrency between transactions.

2.2.1   CORRECTNESS CRITERIA FOR DATABASE TRANSACTIONS

Transactions for computing have their roots in database systems, and so it is natural to examine the extent to which the correctness criteria used for database transactions can be used to develop a sequential semantics for TM. To understand this better, let us revisit the three database properties of atomicity, consistency, and isolation:

Atomicity (specifically, failure atomicity) requires that a transaction execute to completion or, in case of failure, to appear not to have executed at all. An aborted transaction should have no side effects.

Consistency requires that a transaction transform the database from one consistent state to another consistent state. Consistency is a property of a specific data structure, application, or database. It cannot be specified independently of the semantics of a particular system. Enforcing consistency requires a programmer to specify data invariants, typically in the form of predicates. These are typically not specified with TM systems, although some exploration has been made [137].

Isolation requires that execution of a transaction not affect the result of concurrently executing transactions.

What about the "D" *durability* from the full ACID model? One could argue that durability does not apply to TM because volatile memory is inherently not a durable resource, unlike storage on a disk. Conversely, one could argue that the effects of a TM transaction must remain durable so long as the process's memory state is retained; that is, a correct TM should not allow a transaction's effects to be discarded until the process terminates. This is a semantic question of exactly what one takes

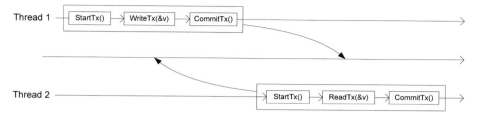

Figure 2.3: Serializability does not have to respect real-time ordering: Thread 1's transaction can appear to run after Thread 2's transaction, even if its execution comes completely before Thread 2's.

"durability" to mean. It indicates once again how care is required in mapping ideas from database transactions onto TM.

2.2.1.1 Serializability
In database systems, the basic correctness condition for concurrent transactions is serializability. It states that the result of executing concurrent transactions on a database must be identical to *a* result in which these transactions executed serially. Serializability allows a programmer to write a transaction in isolation, as if no other transactions were executing in the database system. The system's implementation is free to reorder or interleave transactions, but it must ensure the result of their execution remains serializable.

Although serializability requires that the transactions appear to run in a sequential order, it does not require that this reflects the real-time order in which they run. Figure 2.3 gives an example: Thread 1's transaction can appear to run after Thread 2's transaction, even though Thread 1's transaction executes first. This could be counter-intuitive in a shared-memory programming model— for instance, Thread 1 may complete its transaction, perform some non-transactional synchronization with Thread 2 to announce that it has finished, and then Thread 2's transaction may miss memory updates that were committed by Thread 1.

2.2.1.2 Strict Serializability
Strict serializability is a stronger correctness criteria than ordinary serializability: it requires that if transaction TA completes before transaction TB starts, then TA must occur before TB in the equivalent serial execution. Consequently, for the example in Figure 2.3, strict serializability would require that Thread 1's transaction appears to run before Thread 2's.

This seems a better fit than ordinary serializability, but even strict serializability does not capture all of the requirements we might have for the semantics of TM. In particular, although it specifies legal interactions between transactions, it says nothing about the interaction of transactions with non-transactional code.

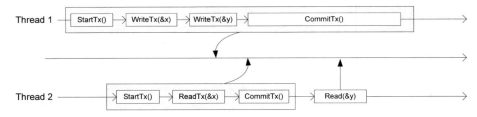

Figure 2.4: Strict serializability does not consider non-transactional accesses, such as the read of y from Thread 2.

Figure 2.4 gives an example: Thread 1's transaction writes to x and to y and is serialized before Thread 2's transaction, which reads from x. Having seen the value in x, Thread 2 might decide to read from y non-transactionally, and expect that it will see the (apparently-previous) write from Thread 1's transaction.

Strict serializability does not require this because it does not consider the role of non-transactional accesses. Indeed, databases do not face this difficulty since they control access to shared data and require all accesses to be mediated by the database. By contrast, programs using TM can execute code outside of any transaction, and they can potentially access data directly as well as transactionally. If transactional and non-transactional code access shared data, their interaction needs to be specified to describe the semantics of transactional memory fully. We return to this kind of interaction in detail in Sections 2.2.3–2.2.5.

2.2.1.3 Linearizability

We introduced the notion of linearizability when considering the behavior of individual operations such as `StartTx`, `ReadTx`, `CommitTx`, etc.: linearizability requires that each of these operations appears to execute atomically at some point between when it is invoked and when it completes.

Linearizability can be used in a similar way to define the semantics of complete transactions. In this case, a transaction would be considered as a single operation, extending from the beginning of its `StartTx` call until the completion of its final `CommitTx`. Linearizability would then require that each transaction appears to take place atomically at some point during this interval. This model can readily accommodate non-transactional read and write operations: as before, these must occur atomically at some point during the read or write's individual execution.

Under linearizability, the example of Figure 2.4 would be guaranteed that if Thread 1's transaction appears to execute before Thread 2's transaction, then Thread 2's non-transactional read would be guaranteed to see Thread 1's transactional write.

This form of linearizability, therefore, looks like an attractive property for a TM system to satisfy. However, it is unclear how this model should accommodate transactions that abort due to conflicts—after all, if each transaction appears to execute atomically at a single instant, then conflicts between transactions will not occur.

2.2.1.4 Snapshot Isolation

Many databases support forms of isolation that are *weaker* than serializability. In databases, these can allow greater concurrency between transactions, by allowing non-serializable executions to commit. They can also simplify the database's implementation by reducing the kinds of conflicts that need to be detected. Researchers have explored whether or not similar correctness criteria can be exploited by TM [263].

Snapshot isolation (SI) is a weaker criteria that allows the reads that a transaction performs to be serialized before the transaction's writes. The reads must collectively see a valid snapshot of memory, but SI allows concurrent transactions to see that same snapshot and then to commit separate sets of updates that conflict with the snapshot but not with one another. For instance, consider this contrived example[1]:

```
// Thread 1                        // Thread 2
do {                               do {
  StartTx();                         StartTx();
  int tmp_x = ReadTx(&x);            int tmp_x = ReadTx(&x);
  int tmp_y = ReadTx(&y);            int tmp_y = ReadTx(&y);
  WriteTx(&x, tmp_x+tmp_y+1);        WriteTx(&y, tmp_x+tmp_y+1);
} while (!CommitTx());             } while (!CommitTx());
```

In this example, the two threads are doing identical work, except that Thread 1 writes the total of x+y+1 to x while Thread 2 writes the total to y. If x==0 and y==0 initially, then under serializability the possible results are x==1 and y==2, or x==2 and y==1, depending on how the transactions occur in the serial order. Snapshot isolation admits a third result of x==1 and y==1, in which both transactions see the same initial snapshot and then commit their separate updates.

It is not currently clear if weaker models such as SI are as useful with TM as they are with databases. First, the semantics of SI may seem unexpected to programmers when compared with simpler models based on serial ordering of complete transactions. Second, implementations of SI for TM do not often seem to offer performance advantages when compared with models such as strict serializability; consequently, performance benefits must come from increased concurrency between transactions, rather than from simplifications within the TM itself. Of course, this may change as research continues; for instance, faster implementations may be developed, or it may be that SI is attractive in distributed settings (Section 4.7) where conflict detection can be more costly than between threads on a chip multiprocessor.

Riegel *et al.* defined an additional correctness criteria, *z-linearizability*. As with classical linearizability, this operates at a whole-transaction basis. However, z-linearizability aims to permit greater concurrency in workloads that mix long-running and short-running transactions [260]. It requires that the set of all transactions is serializable, but that specified sub-sets of transactions are linearizable—long-running transactions might form one such set, and short-running transactions might form another.

---

[1]Throughout this book, unless stated otherwise, variables are assumed to be ordinary non-`volatile` 0-initialized integers.

2.2.2 CONSISTENCY DURING TRANSACTIONS

As we illustrated in the previous section, correctness criteria from databases provide some intuition for the semantics of TM systems, but they do not consider all of the complexities. There are two main areas where they fall short:

First, they specify how *committed* transactions behave, but they do not define what happens while a transaction runs: for instance, is it permitted for a transaction's ReadTx operations to return arbitrary values, so long as the transaction ultimately aborts?

Second, criteria such as serializability assume that the database mediates on *all* access to data, and so they do not consider cases where data is sometimes accessed via transactions and sometimes accessed directly.

A definition for TM's semantics needs to consider both of these areas. Several different approaches have been explored in the literature, and so our aim, in this book, is to survey the options and to examine their similarities and differences. There is not a consensus in the research community on a single "correct" choice; indeed, initial work suggests that perhaps different approaches are appropriate for different kinds of use of TM.

In this initial section, we focus on a transaction-only workload and use this to examine the behavior of the ReadTx operation while a transaction is running. Then, in Sections 2.2.3–2.2.5, we expand the scope to include non-transactional memory accesses as well as transactional ones. This introduces additional complexity, so we try to separate those issues from the simpler problems that occur when only transactions are involved.

2.2.2.1 Inconsistent Reads and Incremental Validation

As we discussed previously, strict serializability provides an intuitive model for the execution of *committed* transactions. However, it does not provide a definition of how transactions should behave when they fail to commit—either because they abort or because they do not get as far as trying to call CommitTx. For instance, consider the following two threads (as usual, x==y==0 initially):

```
// Thread 1                       // Thread 2
do {
  StartTx();
  int tmp_1 = ReadTx(&x);
                                  do {
                                    StartTx();
                                    WriteTx(&x, 10);
                                    WriteTx(&y, 10);
                                  } while (!CommitTx());
  int tmp_2 = ReadTx(&y);
  while (tmp_1 != tmp_2) { }
} while (!CommitTx());
```

Suppose Thread 1 reads from x, and then Thread 2 runs in its entirety. With eager version management and lazy conflict detection, when Thread 1 resumes, it will read Thread 2's update to y and consequently see tmp_1==0, and tmp_2==10. This will cause it to loop forever. This example illustrates that (at least with this kind of TM) doing commit-time validation is *insufficient* to prevent Thread 1 from looping: the thread will never try to commit its transaction.

A transaction such as Thread 1's, which has become inconsistent but not yet detected, the conflict is said to be a *zombie transaction* [83] (also known as a *doomed transaction* [278]). Zombie transactions can cause particular problems in unsafe languages, such as C or C++, because inconsistency can cause incorrect behavior without raising an exception. For instance, the values of `tmp_1` and `tmp_2` could be used as inputs to pointer arithmetic to compute an address to access. If the values are inconsistent then the transaction may try to read from an invalid memory location (causing an access violation); even worse, a write based on a computed address might access a location which is not intended to be used transactionally. Inconsistency can be a problem even in a type-safe language—for instance, Spear *et al.* identified that a zombie transaction may attempt to access a (type-correct) location that is not meant to be used transactionally. How should these problems be handled?

One option is to make the programmer responsible, requiring that they anticipate the anomalies that might be caused by inconsistent reads. For instance, the programmer may need to add explicit validation to loops to guarantee that invalidity is detected and to perform additional validation after reads to detect that a set of addresses that will be used in address arithmetic is consistent.

This "incremental validation" approach is not always satisfactory. The additional validation work can impose a cost; in some STM systems, validating n locations requires n memory accesses, and so ensuring that a series of accesses is valid after each access requires $O(n^2)$ work. Furthermore, this approach couples the program to the particular TM that it is using—e.g., a TM using eager updates allows a zombie transaction's effects to become visible to other transactions, while a TM using lazy updates only allows the effects of committed transactions to become visible. Different anomalies will occur on different TM implementations, making it hard to write portable programs.

2.2.2.2 Opacity

An alternative is for the TM to provide stronger guarantees about the consistency of the values read by a transaction. Guerraoui and Kapałka formalized this as the notion of *opacity* [125], by defining a form of strict serializability in which running and aborted transactions must also appear in the serial order (albeit without their effects being exposed to other threads). A TM supporting opacity must ensure that a transaction's read-set remains consistent during its execution; otherwise, the tentative work could not be part of the serial order because some of the work would have to appear before a conflicting update from another transaction, and some of the work would have to appear after.

Knowing that the TM supports opacity would guarantee that our example will not loop: Thread 1's reads from x and y must form a consistent view of memory, and so the two values must be equal.

Several TM implementations provide opacity (including some from before the definition was crisply formalized). For instance, the TL2 STM (Section 4.3) uses a global clock to identify when data has been updated and requires that all of the data read by a given transaction is consistent at a given timestamp. HTMs typically either provide opacity or guarantee that the effects of zombie transactions will be completely sandboxed (say, in a local cache) and that a transaction cannot loop endlessly while invalid.

Imbs *et al.* [163] introduced a spectrum of consistency conditions in which opacity and serializability are two extreme positions, along with a *virtual world* consistency condition which is intermediate between them.

2.2.3   PROBLEMS WITH MIXED-MODE ACCESSES

The second main difference between database correctness criteria and those for TM is that a semantics for TM must consider the interaction between transactional and non-transactional access to the same data. So far, we have assumed that the memory used via transactions is completely separate from the memory used via non-transactional accesses, and so these concerns have not occurred in our initial examples.

The main problem is that many TMs *do not* provide any conflict detection between transactional and non-transactional accesses, and so programs that involve such conflicts can behave unexpectedly—particularly when zombie transactions are present. These problems occur primarily with STM systems because conflict detection in software relies on support from both of the threads involved: unless non-transactional accesses are modified to play their part, then conflicts involving them will go undetected.

2.2.3.1   Weak and Strong Isolation

Blundell *et al.* [35; 36] introduced the terms *weak atomicity* and *strong atomicity*. Weak atomicity guarantees transactional semantics only among transactions. Strong atomicity also guarantees transactional semantics between transactions and non-transactional code.

The terms "weak/strong atomicity" and "weak/strong isolation" are both used in the literature. They are synonymous, and arguments can be made for either of them to be appropriate. The "isolation" variants can be explained by reference to database terminology where the problem that occurs is that the effects of a transaction are not being isolated correctly from the effects outside the transaction. The "atomicity" variants can be explained with reference to programming language terminology, viewing the work inside the transaction as a single atomic transition—e.g., in an operational semantics for a language. In that context, interleaving of non-transactional work would be a loss of atomicity.

In this book, we use the "isolation" terms because we discuss the links with database transactions at several points through the text.

2.2.3.2   Problems with Weak Isolation

We now examine a series of problems that occur when using a TM with weak isolation. The first set of problems involve unambiguous data races between transactional and non-transactional code. The second set involve *granularity* problems that occur when the data managed by the TM implementation is coarser than the program variables being accessed. The third set of problems involve accesses by zombie transactions in TMs that do not provide opacity. Finally, the last set involve *privatization* and *publication* idioms where the programmer attempts to use transactional accesses to one piece of data to control whether or not another piece of data is shared. As we illustrates, this last set is problematic whether or not opacity is provided.

Many more examples have been identified and studied in the literature, and we do not attempt to reproduce all of them here. Our examples are based on work by Blundell *et al.* [35; 36], Shpeisman *et al.* [288], Grossman *et al.* [115], and Menon *et al.* [219; 220]. In addition, Abadi *et al.* [3] and Moore and Grossman [226] examine how these kinds of examples relate to formal operational semantics for TM.

Examples like these have helped researchers develop an informal understanding of the problems of mixing transactional and non-transactional accesses, but they do not in themselves provide a precise definition of how a TM should behave for all inputs; it is possible that there are insidious cases that examples do not highlight. Therefore, in Section 2.2.5, we discuss methodical approaches to handling these examples, rather than considering them each on a case-by-case basis.

2.2.3.3 Anomalies Common with Lock-Based Synchronization

The first set of problems are reminiscent of those that occur when a lock-based program is not race-free and an asymmetric data race occurs [258]. Shpeisman *et al.* introduced a taxonomy for these problems, identifying a number of common patterns [288]:

- A *non-repeatable read* (NR) can occur if a transaction reads the same variable multiple times, and a non-transaction write is made to it in between. Unless the TM buffers the value seen by the first read, then the transaction will see the update.

- An *intermediate lost update* (ILU) can occur if a non-transactional write interposes in a read-modify-write series executed by a transaction; the non-transactional write can be lost, without being seen by the transactional read.

- An *intermediate dirty read* (IDR) can occur with a TM using eager version management in which a non-transactional read sees an intermediate value written by a transaction, rather than the final, committed value.

2.2.3.4 Anomalies Due to Coarse Write Granularity

Suppose that x and y are two byte-size variables that are located on the same machine word. One might write the following:

```
// Thread 1                          // Thread 2
do {                                 y = 20;
  StartTx();
  tmp_1 = ReadTx(&x);
  WriteTx(&x, 10);
} while (!CommitTx());
```

In this example, a transaction is used to write to x, while y is written directly. This can expose granularity problems if the TM manages data on a per-word or per-cache-line basis, rather than handling these two bytes separately. For instance, if a TM uses lazy version management, the ReadTx operation may fetch the complete word (executing before Thread 2's write and seeing y==0), and the subsequent commit may write back the complete word (executing after Thread 2's write, and storing y==0.

A similar problem can occur with eager version management if Thread 1's transaction records x==0, y==0 in its undo log (before Thread 2's write), and, subsequently, Thread 1 needs to roll back its transaction because of a conflict with another thread.

Shpeisman *et al.* coined the term *granular lost update* (GLU) for this kind of problem. With weak isolation, GLU problems can be avoided by ensuring that the writes made by the TM precisely match the writes made by the program, in this case, by maintaining bitmaps of which sub-word values are actually written or by requiring that transactional and non-transactional data is placed on different words.

2.2.3.5 Anomalies Due to Zombie Transactions

The third kind of problem with weak isolation can be blamed on zombie transactions. Consider the following example, a slight variant of the earlier one in which a zombie transaction looped forever:

```
// Thread 1                          // Thread 2
do {
  StartTx();
  int tmp_1 = ReadTx(&x);
                                     do {
                                       StartTx();
                                       WriteTx(&x, 10);
                                       WriteTx(&y, 10);
                                     } while (!CommitTx());
  int tmp_2 = ReadTx(&y);
  if  (tmp_1 != tmp_2) {
    WriteTx(&z, 10);
  }
} while (!CommitTx());
```

In this variant of the example, Thread 1 only tries to write to z if it sees different values in x and y. A programmer may hope to reason that this write will never occur—either Thread 1's transaction executes first, and sees both variables 0, or Thread 2's transaction executes first and so Thread 1 sees both values 10. However, if Thread 1's transaction can see an inconsistent view of the two variables, then it may attempt to write to z.

This is a serious problem when using eager version management; a concurrent, non-transactional read from z would see the value 10 be written by Thread 1. Shpeisman *et al.* introduced the term *speculative dirty read* (SDR) for this problem. A related *speculative lost update* (SLU) problem occurs if the transactional write to z conflicts with a non-transactional write: the non-transactional write can be lost when the transaction is rolled back.

2.2.3.6 Anomalies Due to Privatization and Publication

There are other, more complex, programming idioms that involve transactions going awry. Variants of these can occur even with TM systems that provide opacity. Consider the following "privatization" example:

```
// Thread 1                    // Thread 2
do {                           do {
   StartTx();                     StartTx();
   WriteTx(&x_priv, 1);           if (ReadTx(&x_priv) == 0)
} while (!CommitTx());              WriteTx(&x, 200);
x = 100;                       } while (!CommitTx());
```

In this example, there are two variables x_priv and x, and the code is written so that if x_priv is 1 then x is meant to be private to Thread 1. Otherwise, x is meant to be shared between the threads. The problem here is more complex than our previous example with a zombie transaction writing to z because, in this case, the question of whether or not x should be used transactionally depends on the value in another variable.

A programmer might hope to reason either that Thread 1's transaction commits first, in which case, Thread 2 will not write to x, or that Thread 2's transaction commits first, in which case, Thread 1's write of 100 will come after Thread 2's write of 200. In either case, the result should be x==100. In addition, a programmer might think that this privatization idiom is reasonable because the code would be correct if the StartTx/CommitTx operations were replaced by acquiring and releasing a lock.

In practice, however, this reasoning is incorrect on many simple TM systems. For instance, consider an implementation using weak isolation, commit-time conflict detection, and eager version management:

- Thread 2 starts its transaction, reads that x_priv is 0, records x==0 in its undo log, and writes 200 to x.

- Thread 1's transaction executes successfully, committing its update to x_priv. At this point, Thread 1's code believes it has private access to x.

- Thread 1 writes 100 into x.

- Thread 2's transaction fails validation, and so the undo log is used for roll-back, restoring x to 0.

Variants of this problem can occur with many TM implementations, and they are not specific to the use of eager version management. The analogous problem with lazy version management is that Thread 2's transaction may be serialized before Thread 1's transaction, but the *implementation* of Thread 2's CommitTx operation may still be writing values back to memory even though Thread 1's CommitTx has finished. This lets the writes of 200 and 100 race, without any guarantee that the "correct" answer of 100 will prevail.

More broadly, this is known as a *delayed cleanup problem* in which transactional writes interfere with non-transactional accesses to locations that are believed to be private. This interference can happen either because a committed transaction has yet to finish writing its updates to memory, or because an aborted transaction has yet to undo its own updates. Shpeisman *et al.* introduced the term *memory inconsistency* for problems in which a transaction commits values to different locations in an arbitrary order.

By analogy with the privatization example, a similar problem can occur with a "racy publication" idiom where one variable's value is intended to indicate whether or not another variable has been initialized. Consider the following example, in which Thread 1 tries to publish a value in x, but Thread 2's transaction reads from x before it checks the publication flag:

```
// Thread 1                      // Thread 2
                                 do {
                                   StartTx();
                                   int tmp = ReadTx(&x);

x = 42;
do {
  StartTx();
  WriteTx(&x_published, 1);
} while (!CommitTx());

                                   if (ReadTx(&x_published)) {
                                     // Use tmp
                                 } } while (!CommitTx());
```

The example is called "racy" because Thread 1's write to x is not synchronized with Thread 2's read from x. Nevertheless, a programmer familiar with mutual exclusion locks might hope that if Thread 2 sees x_published set to 1 then Thread 2 will also see x==42. This would be guaranteed by the Java memory model [207]. However, when using TM, if Thread 2's read from x executes before Thread 1's non-transactional write, then Thread 2 may miss the value being published.

Shpeisman *et al.* identified a related *granular inconsistent read* (GIR) problem that can occur if a TM implementation performs internal buffering: suppose that Thread 2's transaction also reads from a location y, which shares a memory word with x. If the read from y fetches both variables into a per-transaction buffer, then this may lead to the read from x appearing to occur early.

2.2.3.7 Discussion

The examples in this section have illustrated a wide range of anomalies between TMs that provide strong isolation and TMs with weak isolation. One might argue that, given that there are so many problematic examples, we should simply dispense with weak isolation and instead provide strong. This is clearly very attractive from the point of view of defining TM semantics. Many HTM implementations provide strong isolation, and there has been substantial progress in developing STM implementations [5; 155; 281; 288].

However, simply providing strong isolation does not avoid all of the problems that emerge when programming with TM—even if transactions are individually atomic, the programmer must still grapple with the language's broader memory model. For example, consider this idiom which a programmer might attempt to use for thread-safe initialization:

```
// Thread 1                    // Thread 2
do {                           int tmp_1 = ready;
  StartTx();                   int tmp_2 = data;
  WriteTx(&data, 42);          if (tmp_1 == 1) {
  WriteTx(&ready, 1);            // Use tmp_2
} while (!CommitTx());         }
```

A programmer cannot assume that strong isolation guarantees that if Thread 2 sees `ready==1` then it must also see `data==42`. This line of reasoning is only correct if Thread 2's implementation is guaranteed to read from `ready` before it reads from `data`. This ordering is not enforced by many programming languages (e.g., that of Java [207]) or by many processors' weak memory models (e.g., that of the Alpha [70]). In this case, we may need to mark both `ready` and `data` as `volatile` variables, in addition to accessing them transactionally in Thread 1.

Problems like this can be subtle, suggesting that precise definitions are necessary—both for the guarantees made by a TM system, and for the guarantees provided by the language in which transactions are being used.

2.2.4   HANDLING MIXED-MODE ACCESSES: LOCK-BASED MODELS

In the previous section, we used examples to illustrate the kinds of problem that can emerge when a program mixes transactional and non-transactional accesses. Examples can highlight some of the potential problems, but they do not provide a complete approach to defining the semantics for TM, and they do not provide confidence that we have considered the problem from all possible angles. Even if a TM handles specific privatization and publication idioms, then it is possible that there are further examples – even more complicated – which have not yet been identified.

In this section, and the next, we consider two methodical approaches to these problems. First, we focus on *lock-based* models for semantics for TM. These models are defined by relating the behavior of a program using transactions to the behavior of a program using locks. In effect, the behavior of the lock-based implementation forms a reference model for the behavior of the program using TM. The second approach, in Section 2.2.5, is to define the semantics of transactions directly, rather than by relation to existing constructs.

2.2.4.1 Single-Lock Atomicity for Transactions

A basic, pragmatic model for defining transactions is *single-lock atomicity* (SLA), in which a program executes *as if* all transactions acquire a single, program-wide mutual exclusion lock. Intuitively, SLA is attractive because a similarity to basic locking provides evidence that programming with TM is easier than programming using locks. (The term *single global lock atomicity* is sometimes used in the literature; we use single-lock atomicity because the name is shorter).

Unlike correctness criteria from database systems, SLA readily accommodates mixed-mode accesses to data. Going back to the example of strict serializability from Figure 2.4, SLA would require that if Thread 2 sees Thread 1's write to x, then Thread 2 must also see Thread 1's write to y: both transactions must appear to acquire the single lock, and so one must appear to run entirely before the other, or vice-versa. Similarly, under SLA, the privatization idiom from Section 2.2.3.6 is guaranteed to work correctly.

SLA also leads to a definition of what it means for a program using transactions to have a data-race: the transactional program has a data-race if and only if the lock-based program has a data-race. For instance, this contrived example involves a race on x because there is no synchronization between Thread 1's access and Thread 2's access:

```
// Thread 1              // Thread 2
StartTx();               int tmp = x;
WriteTx(&x, 42);
CommitTx();
```

To define the behavior of "racy" examples, one might either have a "catch-fire" semantics (permitting absolutely any catastrophic behavior in the presence of a race), or one might require that the behavior remains consistent with the equivalent lock-based program's behavior (which may itself be "catch-fire"). The emerging C/C++ memory model has catch-fire semantics for races in lock-based programs [43], and so it might be reasonable to follow this approach in programs using transactions. Conversely, the Java memory model gives guarantees about exactly how a racy program can behave [207], and so similar kinds of guarantee may be needed when adding transactions to Java.

Although SLA is attractive, it can be problematic from several points of view. First, it is hard to extend beyond the most basic transactions; TM may provide additional features that are not present with locks—e.g., failure atomicity, condition synchronization, or forms of nesting. It is not clear how to map these features onto a lock-based reference implementation.

Second, the use of SLA as a reference model does not fit with many peoples' intuitive expectations for transactions. Consider the following example, based on one described by Luchangco [202]:

```
// Thread 1              // Thread 2
StartTx();               StartTx();
while (true) { }         int tmp = ReadTx(&x);
CommitTx();              CommitTx();
```

Under SLA, it would be permitted for Thread 1's transaction to start, acquire a global lock, and then enter an endless loop. Thread 2's transaction would then block forever waiting for the global lock.

Programmers might expect that Thread 2's transaction should be able to run immediately, given that it does not conflict with Thread 1's transaction. (A broader question is exactly what kind of progress property should be provided for transactions. We return to this in detail in Section 2.3.1, but the immediate problem with SLA is that the intuition provided by locks may not coincide with the intuition expected for transactions).

Third, SLA may be too strong a correctness criterion. For instance, consider this variant of an "empty publication" example from Menon *et al.* [219]:

```
// Thread 1                      // Thread 2
                                 StartTx();
                                 int tmp = data;

data = 1;
do { StartTx();
} while (!CommitTx())
ready = 1;

                                 if (ready) {
                                   // Use tmp
                                 }
                                 CommitTx();
```

This program has a race on `data` but under SLA in Java or C#, the programmer may nevertheless reason that Thread 1's empty transaction synchronizes on the global lock with Thread 2's transaction, and so if Thread 2 sees `ready==1`, then it must also see `data==1`. Supporting this form of programming introduces synchronization between these two transactions—even though there is no overlap in the data that they access.

2.2.4.2 Disjoint Lock Atomicity (DLA)

To remedy some of these shortcomings, Menon *et al.* introduced a series of lock-based models for TM [219]. These are weaker than SLA, in the sense that they do not support programming idioms that can be used under SLA. Conversely, they can allow for a wider range of implementation techniques which are faster or more scalable than those known for SLA.

The first of these weaker models is *disjoint lock atomicity* (DLA). Under DLA, before each transaction runs, it must acquire a minimal set of locks such that two transactions share a common lock if and only if they conflict. This means that there is no synchronization between transactions that access disjoint data. Consequently, DLA does not support the empty publication idiom. However, it does support programming idioms which involve synchronization on actual data—e.g., the racy publication idiom may be used.

Unlike SLA, it is not generally possible to directly translate a program using DLA transactions into an equivalent lock-based one: DLA is *prescient*, in the sense that the locks can only be acquired given knowledge of the transaction's future data accesses. Nevertheless, DLA relates transactional behavior to the familiar behavior of locking.

2.2.4.3   Asymmetric Lock Atomicity (ALA)

A further weakening of DLA is *asymmetric lock atomicity* (ALA) [219]. Under DLA, a transaction must appear to acquire all of its locks before it starts. ALA relaxes this and allows the acquisition of write locks to be delayed up until the point of the transaction's first write access to a given location. ALA guarantees that a transaction has exclusive access to all of the data that it will read, and so the racy publication idiom is supported.

2.2.4.4   Encounter-time Lock Atomicity (ELA)

Menon *et al.*'s final lock-based model is *encounter-time lock atomicity* (ELA) [219]. This relaxes ALA still further: *all* lock acquisitions may be delayed until the point of first access. This supports the publication idiom only if the compiler avoids traditional optimizations that may hoist a read above a conditional branch.

2.2.5   HANDLING MIXED-MODE ACCESSES: TSC

Lock-based models, such as SLA, DLA, ALA, and ELA, provide a series of alternative semantics for transactions. They provide a methodical way of considering examples such as privatization and publication: rather than looking at the examples on a case-by-case basis, one must ask which locks would be acquired in a given example, and then consider how the resulting lock-based synchronization would behave.

An alternative approach is to define the semantics for TM directly, without this intermediate step of using locks. Direct definitions can accommodate TM-specific extensions for condition synchronization, failure atomicity, nesting, and so on, more readily than a lock-based model.

One way to define TM is to build on the idea of how sequential consistency (SC) provides an ideal semantics for ordinary shared memory [187]. Under SC, threads make reads and writes to shared memory, and these accesses appear to take place instantaneously—without being buffered in caches, or reordered during compilation or by a processor's relaxed memory model. This provides a kind of "gold standard" that one might like an implementation to satisfy.

Dalessandro *et al.* extend SC to a notion of *transactional sequential consistency* (TSC) [76], building on their work on ordering-based semantics for STM [299], and the work of Grossman *et al.* [115]. TSC requires that the effects of a transaction being attempted by one thread are not interleaved with *any* operations being executed by other threads. In particular, note that TSC is a stronger property than simply *strong isolation*. Strong isolation requires that the *isolation* of transactions is complete but does not in itself define how operations may be re-ordered during compilation. Under TSC, no re-ordering may occur. Abadi *et al.* [3] and Moore *et al.* [226] concurrently defined similar notions at the level of programming languages that provide atomic constructs.

Figure 2.5 illustrates TSC: all of the operations from the two threads appear to be interleaved on a common time-line, but the operations from a given transaction must remain together.

In practice, neither SC nor TSC can be supported on all programs by practical implementations. Instead, it is usual to combine either model with a notion of *race-freedom*, and to require that if

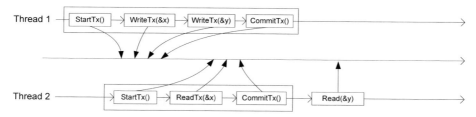

Figure 2.5: Transactional sequential consistency (TSC): the threads' operations are interleaved, with nothing intervening between operations from a given transaction.

a program is race-free then it should be implemented with SC or TSC. As with the handling of data races under SLA, if a program is not race-free, then it may be supported with weaker guarantees, or none at all.

Adve and Hill pioneered this approach with SC, and proposed that the criteria for whether or not a program is "race-free" should be defined in terms of its hypothetical execution on SC [10]. This kind of model is called *programmer centric*; it uses the notion of SC that the programmer is familiar with, rather than the details of a particular processor's memory system. In effect, this forms a contract between the implementation and the programmer: so long as the programmer writes a program that would be race-free under SC, then a real processor must implement it with SC.

Consequently, when designing a transactional programming model using TSC, the question is exactly what set of programs should be considered to be race-free. We shall look at three different definitions that have been studied in the literature: static separation, dynamic separation, and forms of data-race freedom.

2.2.5.1 Static Separation

A *static separation* programming discipline requires that each mutable, shared location is either always accessed transactionally or always accessed non-transactionally. A programmer can achieve this by ensuring that, if a transaction accesses a given location, then either (*i*) the location is immutable, (*ii*) the location is thread-local and the correct thread is accessing it, or (*iii*) the location is never accessed non-transactionally. It is easy to see that any of these conditions suffices to prevent any interference between a transaction and a non-transaction.

Static separation prohibits programming idioms where a mutable shared location changes between being accessed transactionally and being accessed directly: e.g., it prohibits initializing a data structure non-transactionally before it is shared by transactions, or from privatizing a shared data structure and accessing it directly. This means that the privatization and publication idioms from Section 2.2.3.6 do not follow static separation.

For programs that do obey static separation, a TM using weak isolation and a TM with strong isolation are indistinguishable. This provides a lot of flexibility to the TM system's implementer. In addition, the use of a statically-checked type system can provide compile-time checking of whether or not a program is race-free. Abadi *et al.* and Moore and Grossman have investigated this relationship from a formal viewpoint [3; 226].

However, data types can provide a blunt tool for enforcing static separation. For example, code executed in a transaction could be restricted to access data whose type is marked as transactional. It would be a detectable error to access data of this type outside a transaction. This is the approach taken by STM-Haskell [136], where mutable STM locations are accessible only within transactions.

In Haskell, however, most data is immutable and remains accessible both inside and outside a transaction. In non-functional languages, most data is mutable and would need to be partitioned into transactional and non-transactional sections. This distinction effectively divides a program into two worlds, which communicate only through immutable values.

This division complicates the architecture of a program, where data may originate in the non-transactional world, be processed in transactions, and then return to the non-transactional world for further computation or IO. Structuring this kind of program to obey static separation either requires data to be marshaled between separate transactional and non-transactional variants, or for transactions to be used ubiquitously. The first option is cumbersome, and the second may perform poorly [284].

2.2.5.2 Dynamic Separation

Dynamic separation conceptually splits the program heap into transactional and non-transactional parts, but it provides operations to move data between the two sections. This mitigates some of the difficulties of programming with static separation, while still allowing flexibility over a range of TM implementations. For instance, Abadi *et al.* describe an implementation of dynamic separation which is built over a TM that uses lazy conflict detection and eager version management [1; 2]. Intuitively, it operates by adding checks within the TM implementation that all of the locations being accessed are part of the transactional heap; this prevents eager updates made by zombie transactions from conflicting with non-transactional work elsewhere in the application. Variants of dynamic separation can support shared read-only data, as with static separation in STM-Haskell. A disadvantage of dynamic separation, over static, is that it is not so amenable to compile-time checking for whether or not a program is race-free.

Lev and Maessen propose a related idea for dynamically tracking which objects are guaranteed to be local to a given thread and which objects might be shared between multiple threads [189]. Reads and writes to both kinds of objects are performed via accessor methods. These methods perform direct access in the case of local objects and enforce transactional access in the case of shared objects. Each object has a flag indicating whether or not it is shared; objects start local and transition to shared when a global variable or a shared object is updated to point to them. Shpeisman *et al.* use a similar dynamic technique to Lev and Maessen's to improve the performance of an STM system providing strong isolation [288].

2.2.5.3 Transactional Data-Race-Freedom

A stronger programming discipline than dynamic separation is *transactional data-race freedom* (TDRF) [76]. A transactional data-race is one in which a conflict occurs between a transactional access and a non-transactional access. A program is TDRF if it has no ordinary data races, and no transactional data races either. Abadi *et al.* introduced the term *violation* for a transactional data race, and *violation-freedom* for execution where there are no such races [3]. Informally, TDRF is a stronger property than violation freedom because a TDRF program must be free from ordinary data races, as well as just ones involving transactions.

TDRF is defined with reference to that program's behavior under TSC—not with reference to its behavior on any particular TM implementation. Consequently, accesses that might only be attempted by a zombie transaction *do not* qualify as data races, nor do accesses to adjacent bytes in memory (unless the broader definitions of SC for that language or machine would treat sub-word accesses as a race).

The examples from the earlier sections on anomalies due to "write granularity", "zombie transactions", "privatization and publication" are all TDRF. TDRF, therefore, characterizes the sense in which these examples are valid, but other examples (such as "racy publication" and "empty publication") are incorrect.

2.2.5.4 Summary

In Sections 2.2.4 and 2.2.5, we have illustrated two approaches for giving methodical definitions of how a TM system should behave in the presence of transactional and non-transactional accesses to the same locations. The first approach, typified by SLA, is to relate the behavior of a program using transactions to the behavior of a lock-based reference implementation. The second approach, typified by TSC, is to define the semantics of transactions directly by extending the notion of sequential consistency.

There is not yet a consensus on exactly which of these approaches is best. There is, nevertheless, a great deal of common ground between the different models. First, both approaches provide a framework for defining the behavior of programs in general, rather than treating specific examples like "racy publication" on an *ad-hoc* case-by-case basis. A precise definition is important if programs are to be written correctly and to be portable across implementations. The second important similarity is that both approaches introduce a notion of programs being race-free. The set of programs which are race-free under the SLA, DLA, ALA and ELA models is the same from one model to another and corresponds to the notion of TDRF.

2.2.6   NESTING

Having examined the basic semantics for TM, we now consider the additional complications introduced by *nesting*. A nested transaction is a transaction whose execution is properly contained in the dynamic extent of another transaction.

There are many ways in which nested transactions can interact, and the different design choices have been studied independently of TM (e.g., Haines *et al.* investigate a rich system that combines concurrency, multi-threading, and nested transactions [128]). For now, however, we assume that (*i*) each transaction can have at most one pending child transaction within it ("linear nesting" [229]), (*ii*) the inner transaction sees modifications to program state made by the outer transaction, and (*iii*) there is just one thread running within a given transaction at a given time. In this restricted setting, the behavior of the two transactions can still be linked in several ways:

Flattened Nesting.   The simplest approach is *flattened nesting*, also known as *subsumption nesting*. In this case, aborting the inner transaction causes the outer transaction to abort, but committing the inner transaction has no effect until the outer transaction commits. This is reminiscent of re-entrant locks: a lock may be acquired multiple times by the same thread but is only released once all of the matching unlock operations have occurred. The outer transaction sees modifications to program state made by the inner transaction. If the inner transaction is implemented using flattening in the following example, then when the outer transaction terminates, the variable x has value 1:

```
int x = 1;

do {
  StartTx();
  WriteTx(&x, 2);     // Outer write
  do {
    StartTx();
    WriteTx(&x, 3);   // Inner write
    AbortTx();
    ...
```

Flattened transactions are easy to implement, since there is only a single transaction in execution, coupled with a counter to record the nesting depth. However, flattened transactions are a poor programming abstraction that subverts program composition if an explicit `AbortTx` operation is provided: an abort in a transaction used in a library routine must terminate all surrounding transactions.

Closed Nesting.   Transactions that are not flattened have two alternative semantics. A *closed transaction* behaves similarly to flattening, except the inner transaction can abort without terminating its parent transaction. When a closed inner transaction commits or aborts, control passes to its surrounding transaction. If the inner transaction commits, its modifications become visible to the surrounding transaction. However, from the point of view of other threads, these changes only become visible when the outermost transaction commits. In the previous example, if closed nesting were to be used, variable x would be left with the value 2 because the inner transaction's assignment is undone by the abort.

For executions that commit successfully, the behavior of flattening and closed nesting is equivalent. However, closed nesting can have higher overheads than flattened transactions and so, if commits are common, a TM system can use flattening as a performance optimization—i.e., optimistically executing a nest of transactions using flattening, and if an abort occurs in an inner transaction, then re-executing the nest with closed nesting.

Open Nesting.   A further type of nesting is *open nesting*. When an open transaction commits, its changes become visible to all other transactions in the system, even if the surrounding transaction is still executing [231]. Moreover, even if the parent transaction aborts, the results of the nested, open transactions will remain committed. If open nesting is used in the following example then, even after the outer transaction aborts, variable z is left with the value 3:

```
int x = 1;

do {
  StartTx();
  WriteTx(&x, 2);      // Outer write
  do {
    StartTx();
    WriteTx(&z, 3);    // Inner write
  } while (!CommitTx());
  AbortTx();
  ...
```

As we discuss in Section 2.3.5, open nesting is intended as an abstraction to allow higher-level notions of conflict detection and compensation to be built (e.g., allowing two transactions to concurrently insert different items into a hashtable, even if the implementations involves updates to a common location). This can allow greater concurrency between transactions—for instance, allowing concurrent transactions to increment a single shared counter without provoking a conflict. Nevertheless, open transactions do permit a transaction to make permanent modifications to a program's state, which are not rolled back if a surrounding transaction aborts—e.g., writing entries to a debug log, or simply recording how far a transaction has reached in order to identify conflict hot-spots [107].

Open transactions can subvert the isolation of a transaction, and so using them correctly requires great care. For instance, the semantics can be particularly subtle if the inner transaction reads data written by the outer one or if the inner transaction conflicts with subsequent work by the outer transaction. Moss and Hosking describe a reference model for nested transactions that provides a precise definition of open and closed nesting [229].

Zilles and Baugh described an operation similar to open nested transactions which they called "pause" [346]. This operation suspends a transaction and executes non-transactionally until the transaction explicitly resumes. As with open nesting, the semantics of such operations can be subtle; for instance, whether or not an implementation may roll back the enclosing transaction while the "pause" operation is executing. Unlike open nesting, operations in a pause do not have transactional semantics and must use explicit synchronization.

Parallel Nesting.   Appropriate semantics for nested transactions become murkier if, instead of restricting ourselves to linear nesting, we consider models where multiple transactions can execute in parallel within the same parent transaction. There are many design choices in this kind of system.

A key decision is how to handle two computations C_1 and C_2 that occur in parallel within a single transaction; must these be serialized with respect to one another as two nested transactions or can they interact freely? Permitting parallel code to be reused within transactions suggests that they should behave as normal.

Similar questions emerge in combinations of OpenMP and transactions [25; 225; 323]. Volos *et al.* distinguish *shallow* nesting, in which a transaction may have multiple constituent threads, but no further transactions within it, and *deep* nesting in which transactions occur within parallel threads within transactions [323]. They define the semantics using a "hierarchical lock atomicity" (HLA) model, by analogy with SLA. Conceptually, in HLA, one lock is used for synchronization between top-level transactions, and then additional, separate locks are used for threads belonging to the same parent (e.g., threads working on the implementation of the same transactional OpenMP loop). Volos *et al.*'s implementation is designed to support programs that are race-free under this definition. It supports shallow nesting by extending the TM implementation to manage logs from different threads. Deep nesting is supported by using locks following the HLA model (arguing that maintaining the parent/child relationships between threads and transactions would be likely to add a substantial overhead).

Vitek *et al.* [320] formalize a language with nested, parallel transactions as an extension to a core language based on Java. Agrawal *et al.* investigate a programming model that permits parallelism within transactions in an extension of Cilk [13]. Moore and Grossman also examine a model that combines nesting and parallelism [226]. Ramadan and Witchel introduce a kind of nesting that supports various forms of coordination between sibling transactions [252] (e.g., to attempt a set of alternative transactions and to commit whichever completes first).

More broadly, all these different forms of nesting provide another illustration of TM features which cannot be specified very clearly as extensions to a single-lock atomicity model.

2.3   PERFORMANCE, PROGRESS AND PATHOLOGIES

In Section 2.2, we described techniques that have been used to define the semantics of transactions. In this section, we examine TM techniques from the point of view of their performance—both in practical terms of how fast transactions run and in terms of any assurance that a thread running a transaction will eventually be able to complete it.

There are several different angles to TM performance, and it is worthwhile trying to distinguish them:

Inherent Concurrency.   The first of these is the inherent concurrency available in a workload using transactions: what is the optimal way in which a TM system could schedule the transactions, while providing specified semantics (e.g., TSC)? This provides a bound on the possible behavior that a practical TM implementation could provide. Some workloads provide a high degree of inherent concurrency—e.g., if all of the transactions access different pieces of data. Other workloads provide a low degree of concurrency—e.g., if every transaction is performing a read-modify-write operation on a single data item.

Actual Concurrency.   The second consideration is the extent to which the concurrency obtained by a TM implementation compares with the inherent concurrency that is available. The granularity at which a TM performs conflict detection will affect how the actual concurrency compares with the inherent concurrency available in a workload. For instance, a TM using a single lock to implement SLA would perform very poorly under this metric for most workloads.

Guarantees.   The third consideration is whether or not the TM implementation gives any *guarantees* about the progress of transactions, rather than just trying to run them "as fast as possible". For instance, is there any kind of fairness guarantee about how the transactions being run by different threads will behave?

Sequential Overhead.   The final consideration is the sequential overhead of a TM implementation: in the absence of any contention, how fast does a transaction run when compared with code without any concurrency control? If a TM system is to be useful, then using it must not incur such a high overhead that the programmer is better off sticking with a sequential program. It is, therefore, important to keep in mind, when considering practical implementations, that overall performance depends not just on the algorithmic design choices which we focus on in this book, but also on the maturity of an implementation, and the use of appropriate implementation techniques.

In this section, we examine the ways in which TM implementations affect the performance that is obtained (or not) in a workload, and on the techniques with which a workload can be structured to increase the concurrency that is available.

We start by introducing the terminology of *nonblocking synchronization*, which provides a framework for defining the kinds of progress guarantees that are offered by many TM systems (Section 2.3.1). We then discuss the interaction between the TM's conflict detection mechanisms and the way in which it executes different workloads (Section 2.3.2), and the use of *contention management* techniques and *transaction scheduling* to improve the practical performance of programs (Section 2.3.3). We describe the techniques used in TM systems to reduce conflicts between transactions (Section 2.3.4). Finally, we show how the inherent concurrency in workloads can be increased by incorporating higher-level notions of conflicts, rather than working at the level of individual reads and writes (Section 2.3.5).

2.3.1   PROGRESS GUARANTEES

Here is an example signaling protocol between two transactions. If x is initially 0, then Thread 1's transaction sets the variable to 1, and Thread 2's transaction loops until it sees a non-zero value:

```
// Thread 1                      // Thread 2
do {                             do {
  StartTx();                       StartTx();
  WriteTx(&x, 1);                  int tmp_1 = ReadTx(&x);
} while (!CommitTx());             while (tmp_1 == 0) { }
                                 } while (!CommitTx());
```

How should this example behave? Intuitively, one might like this idiom to be a correct way of synchronizing between the two threads, and in many cases, it does seem to work: if Thread 1 executes first, then Thread 2 will see the update to x, and commit its transaction. This happens irrespective of the implementation techniques used by the TM.

The problem is when Thread 2 starts its transaction *before* Thread 1's. In this case, many implementations can leave Thread 2 spinning endlessly:

- An implementation using pessimistic concurrency control might prevent the two conflicting transactions from proceeding concurrently. Therefore, Thread 2's transaction will prevent Thread 1's transaction from executing, and so Thread 2 will wait forever.

- An implementation using optimistic concurrency control might leave Thread 2 spinning if the TM system only uses commit-time conflict detection.

If programmers are to use transactions, then they need precise guarantees of whether or not this kind of program is required to make progress. Conversely, the language implementer needs a precise definition of how TM should behave if they are to be confident in designing a correct implementation. Defining liveness properties for TM is an open research question; liveness properties are often subtle, and explicit guarantees about liveness are often omitted from specifications of synchronization constructs, programming languages, and processors.

A possible starting point are concepts from *nonblocking synchronization*. These provide a framework for defining one aspect of the kinds of progress that a TM system might provide. As with the semantics of transactions (Section 2.2, and Figure 2.2), we are concerned with progress at two levels:

- *TM-level progress* of the individual operations such as StartTx and CommitTx—e.g., if lots of threads are trying to commit transactions at the same time, then can the TM implementation livelock within the CommitTx operation?

- *Transaction-level progress* of complete transactions through to a successful commit: e.g., if one thread is executing transaction TA while another thread is executing transaction TB, then is it guaranteed that at least one of the transactions will commit successfully, or can they both get in one another's way, and both abort?

At either of these two levels, a *nonblocking* algorithm guarantees that if one thread is pre-empted mid-way through an operation/transaction, then it cannot prevent other threads from being able to make progress. Consequently, this precludes the use of locks in an implementation, because if the lock-holder is pre-empted, then no other thread may acquire the lock. Three different nonblocking progress guarantees are commonly studied:

The strongest of these guarantees is *wait-freedom* which, informally, requires that a thread makes forward progress on its own work if it continues executing [141]. At the TM-level, "forward progress" would mean completion of the individual TM operations that the thread is attempting. At the transaction level, forward progress would mean execution of transactions through to a successful commit. In either case, wait-freedom is clearly a very strong guarantee because the thread is required to make progress no matter what the other threads in the system are trying to do. Although there are general-purpose techniques to build wait-free algorithms, the resulting implementations are usually slow and only applied in niche settings.

The second nonblocking progress guarantee is *lock-freedom*. This requires that, if any given thread continues executing then *some* thread makes forward progress with its work. This is weaker than wait-freedom because it doesn't guarantee that if Thread 1 continues executing then Thread 1 itself will make progress. However, lock-freedom precludes livelock, and the use of actual locking in an implementation. Lock-free systems usually involve a *helping* mechanism, so that if one thread finds that it cannot make progress with its own work, then it will help whichever thread is in its way.

The third nonblocking progress guarantee is *obstruction-freedom* [145]. This requires that a thread can make progress with its own work if other threads do not run at the same time. Intuitively, a thread that has been preempted cannot obstruct the work of a thread that is currently running. This precludes the use of ordinary locks because if thread T1 were to be pre-empted while holding a lock that thread T2 needs, then T2 would have no option but to wait. Obstruction-free algorithms do not, in themselves, prevent livelock—T1 and T2 may continually prevent one another from making progress if both threads are running. To remedy this, obstruction-free algorithms are usually combined with a contention manager that is responsible for causing T1 or T2 to back off temporarily so that both operations can complete; in effect, the contention manager is responsible for ensuring liveness, while the underlying obstruction-free algorithm is responsible for correctness [145].

From a practical point of view, many of the earliest STM implementations supported non-blocking TM operations—usually with obstruction-freedom or lock-freedom. We return to these STM systems and the implementation techniques that they use in Section 4.5.

However, whether or not a TM system needs to be nonblocking has remained a contentious point. Ennals suggested that it is unnecessary to provide nonblocking progress guarantees in a TM implementation [100]—arguing that a modern programming language runtime system controls the scheduling of application threads, and so the problem of one thread being pre-empted while holding a lock is not really a concern. Many researchers developed lock-based STM systems which aimed to be faster than nonblocking designs (e.g.,[83; 85; 99; 138; 274]).

Conversely, recent STM designs have demonstrated that nonblocking progress guarantees can be provided with performance comparable to lock-based designs [208; 313; 314]. The authors of these systems point out that some potential uses of TM *require* the use of nonblocking synchronization—

for example, if transactions are used for synchronization between OS code and an interrupt handler, then a deadlock would occur if the interrupt handler were to wait for code that it has interrupted.

In addition to nonblocking progress guarantees, other notions of liveness are useful in TM systems. Guerraoui *et al.* introduced the notion of a system-wide *pending commit* property, meaning that if a set of transactions are executing concurrently, then at least one of them will be able to run to completion and commit [123]. A TM which satisfies this property cannot abort all of the transactions involved in a conflict.

Some of the trade offs between progress guarantees and TM design have been investigated from a formal point of view; for instance Attiya *et al.* showed that an implementation of TM cannot provide opacity, disjoint-access parallelism [164], and also allow wait-free read-only transactions [23]. Guerraoui and Kapałka examine obstruction-free transactions from a formal viewpoint, showing that obstruction-free TMs cannot be implemented from ordinary read/write operations, but that they can be built from primitives which are strictly less powerful than atomic compare-and-swap [124].

Guerraoui *et al.* formalized an additional aspect of progress with the notion of *permissiveness* [119]. Informally, a TM is permissive if it does not abort transactions unnecessarily; for instance, a TM that provides opacity should not abort transaction schedules, which are opaque. Existing STM implementations are usually not permissive. In many cases, this is because two transactions can execute their read/write operations in a consistent order (transaction TA always before TB), but they invoke their commit operations in the opposite order (transaction TB first, then TA). Extra bookkeeping is often needed to allow this reversed commit order, and the cost of this bookkeeping can outweigh the gain in concurrency. Guerraoui *et al.* also introduce a *strong progressiveness* progress property that is met by some practical TM implementations. A TM provides strong progressiveness if (*i*) a transaction that runs without conflict is always able to commit, and (*ii*) if a number of transactions have "simple" conflicts on a single variable, then at least one of them will commit.

2.3.2 CONFLICT DETECTION AND PERFORMANCE

The high-level design choices between different conflict detection mechanisms can have a substantial impact on the performance of a TM system and the degree to which it is suited to different kinds of workload.

There is a recurring tension between design choices that avoid *wasted work* (where a transaction executes work that is eventually aborted) and design choices that avoid *lost concurrency* (where a transaction is stalled or aborted, even though it would eventually commit). To illustrate this, let us consider a synthetic workload that can exhibit both problems:

```
// Thread 1                      // Thread 2
do {                             do {
  StartTx();                       StartTx();
  WriteTx(&x, 1);                  WriteTx(&x, 1);
  // Long computation              // Long computation
  ...                              ...
  if (Prob(p)) {                   if (Prob(p)) {
    AbortTx();                       AbortTx();
} } while (!CommitTx());         } } while (!CommitTx());
```

In this example, the threads are doing identical work, and both make conflicting writes to x before starting a long computation. Once the computation is complete, they abort with probability p. Otherwise, they attempt to commit. For this example, we shall assume that there are no other transactions running in the system and that an explicit AbortTx() operation is used to trigger an abort; in practice, these aborts may be due to the code performing an operation that frequently introduces a conflict.

If the probability p is high, then most of the attempted transactions will call AbortTx. Consequently, it is worthwhile executing both thread's transactions concurrently, in the hope that one of them will choose to commit. Conversely, if the probability p is low, then it is best to run only one of the transactions at a time, because the conflicts on x will cause the other transaction to abort, wasting the computation it has done.

Let us consider how this example would behave with various conflict detection mechanisms:

• *Eager detection of conflicts between running transactions:* in this case, a conflict is detected between the two accesses to x. This occurs even if the value of p is high, and, therefore, even when it is unlikely that either transaction would actually try to commit. Contention management or transaction-aware scheduling is needed to avoid livelock (Section 2.3.3).

• *Eager detection of conflicts, but only against committed transactions:* in this case, the transactions can proceed concurrently, and a conflict is detected when one of them tries to commit. Livelock can be prevented by a simple "committer wins" conflict resolution policy. If p is high then this will behave well. If p is low, then one thread's work will be wasted. Nevertheless, the system has the pending-commit property because at any instant one or other of the threads is doing work that will be committed [123].

• *Lazy detection of conflicts between running transactions:* in this case, a conflict occurs between the two threads' accesses to x, but the transactions are allowed to continue nevertheless. The resulting behavior will depend on how the tentative conflict is resolved when one of the transactions tries to commit (e.g., whether the system attempts to enforce some form of per-thread fairness).

• *Lazy detection of conflicts, but only against committed transactions:* in this case, one transaction may continue running even though another has committed a conflicting update. Once such a commit has happened, the first transaction is doomed to abort and will be wasting its work.

It is clear that eager detection of tentative conflicts puts a particular strain on mechanisms to avoid repeated conflicts between a set of transactions. As Spear *et al.* [300], Shriraman *et al.* [290], and Tomic *et al.* [318] observe, such systems effectively require that the TM anticipates which of the transactions involved in a conflict is most likely to be able to commit in the future. In a system with a non-trivial number of conflicts, this is difficult to do well because either transaction might go on to abort because of a conflict with some other transaction.

To avoid these problems, Welc *et al.* designed a hybrid conflict detection system [332]. Under low contention, it detects conflicts eagerly and uses eager version management. Under high contention, it switches to allow lazy conflict detection and uses lazy version management.

Shriraman *et al.* and Spear *et al.* also argued that mixed forms of conflict detection are effective in practice [290; 291; 300]: performing eager detection of conflicts between concurrent writers but performing lazy detection of read-write conflicts against committed transactions. They observe that this provides most of the concurrency benefits of using full lazy conflict detection, while avoiding much of the wasted work.

Bobba *et al.* introduced a taxonomy of "performance pathologies" that occur with many TM implementations [40]:

- *Friendly fire.* As in our previous example, a transaction TA can encounter a conflict with transaction TB and force TB to abort. However, if TB is immediately re-executed, then it might conflict with TA and force TA to abort. Without contention management, this cycle can repeat.

- *Starving writer.* This occurs with TMs that give priority to readers: a writer TA can stall, waiting for a reader TB to complete, but another reader TC starts before TB is finished. A set of readers, continually starting and committing, will prevent TA from ever making progress.

- *Serialized commit.* Some TM implementations involve short, global critical sections to serialize transactions. If the transactions themselves are short, then the serialization introduced by these critical sections can harm scalability.

- *Futile stalling.* Futile stalling can occur if a transaction TA encounters a conflict with transaction TB, and TA decides to wait until TB has finished. Intuitively, this makes sense because it allows TA to continue later, rather than aborting. However, if TB aborts then TA has been delayed without benefit.

- *Starving elder.* TMs with lazy conflict detection and a committer-wins policy can prevent a long-running transaction from being able to commit: by the time it tries to commit, it has conflicted with a more recent short-running transaction.

- *Restart convoy.* TMs with lazy conflict detection can form "convoys" comprising sets of transactions making similar memory accesses. At most, one transaction can commit from each such convoy, with the remaining transactions needing to be re-executed.

- *Dueling upgrades.* This pathology occurs when two transactions read from a location and then subsequently update it. With eager conflict detection, at most one of the duelists can be permitted to upgrade.

In addition, with HTM, care is needed to provide performance isolation between different processes. Typical hardware does not provide complete performance isolation in any case (for instance, hardware threads may compete for space in a cache), but TM introduces particular problems because a process cares about the rate at which its complete transactions commit, as well as the rate at which their constituent instructions execute.

Zilles and Flint identify specific problems that occur with some HTM designs [347]. Some TMs require a complete transaction to be present in *physical* memory if it is to commit. Unlike traditional virtual memory systems, this is a hard requirement: the transaction cannot run at all unless it has the required memory. Mechanisms for handling large transactions can be problematic, if they require transactions to be serialized in their execution, or if they involve "overflow" modes which change system-wide execution, or if they use conflict detection mechanisms that may generate false positives between concurrent transactions from different processes. Finally, if an HTM system were to allow a transaction to cross from user-mode into kernel-mode, then conflicts could occur between kernel-mode execution on behalf of one process and kernel-mode execution on behalf of a different process.

2.3.3   CONTENTION MANAGEMENT AND SCHEDULING

One way to mitigate poor performance caused by conflicts between transactions is to use a *contention manager* (CM), which implements one or more *contention resolution policies*. When a conflict occurs, these policies select whether to abort the "acquiring" transaction that detects the conflict, the "victim" transaction that it encounters, and whether or not to delay either transaction. Many different contention management policies have been studied, varying considerably in complexity and sophistication [122; 123; 278; 280]:

- *Passive.* This is the simplest policy: the acquiring transaction aborts itself and re-executes.

- *Polite.* The acquiring transaction uses exponential backoff to delay for a fixed number of exponentially growing intervals before aborting the other transaction. After each interval, the transaction checks if the other transaction has finished with the data.

- *Karma.* This manager uses a count of the number of locations that a transaction has accessed (cumulative, across all of its aborts and re-executions) as a priority. An acquiring transaction immediately aborts another transaction with lower priority. However, if the acquiring transaction's priority is lower, it backs off and tries to acquire access to the data N times, where N is the difference in priorities, before aborting the other transaction.

- *Eruption.* This manager is similar to Karma, except that it adds the blocked transaction's priority to the active transaction's priority; this helps reduce the possibility that a third transaction will subsequently abort the active transaction.

- *Greedy.* The Greedy CM provides provably good performance when compared with an optimal schedule. In the absence of failures, every transaction commits within a bounded time, and the time required to complete n transactions that share s locations is within a factor of $s(s + 1) + 2$. A transaction is given a timestamp when it starts its first attempt. Upon contention, transaction TA aborts transaction TB if TB has a younger timestamp than TA (or if TB is itself already waiting for another transaction).

- *Greedy-FT.* This extends the Greedy policy and provides robustness against failed transactions via a timeout mechanism.

- *Kindergarten.* This manager maintains a "hit list" of transactions to which a given transaction previously deferred. If transaction TA tries to obtain access to data in use by transaction TB, then if TB is on the list, TA immediately terminates it. If TB is not on the list, then TA adds it to the list and then backs off for a fixed interval before aborting itself. This policy ensures that two transactions sharing data take turns aborting.

- *Timestamp.* This manager aborts any transaction that started execution after the acquiring transaction.

- *Published Timestamp.* This manager follows the timestamp policy but also aborts older transactions that appear inactive.

- *Polka.* This is a combination of the Polite and Karma policies. The key change to Karma is to use exponential backoff for the N intervals.

Selecting between contention managers can be dependent on the TM, on the workload, and on the form of concurrency control used by the TM. For instance, as shown by the "friendly-fire" example in the previous section, if a TM uses eager conflict detection, then the CM must get one or other of the conflicting transactions to back off sufficiently long for the other to complete.

No policy has been found which performs universally best in all settings, though Polka has often worked well [280], as have hybrids that switch from a simple passive algorithm to a more complex one if contention is seen, or if transactions become long-running [93].

Work such as that of Shriraman *et al.* and Spear *et al.* has shown that if conflicts are only detected with *committed* transactions, then the choice of CM algorithm is much less critical because the underlying TM system ensures progress for more workloads [290; 291; 300].

In addition, a further challenge in contention management is how to combine CM algorithms *within* the same program; many early TM benchmarks exhibit a broadly homogeneous mix of transactions, and so a single strategy is appropriate across all of the threads. In more complex workloads, there may be sections of the program that favor different strategies—either varying across time or concurrently for different kinds of data.

2.3.3.1   Contention Management in Real-Time Systems

Pizlo *et al.* describe an application of transactional memory to resolve priority inversion problems involving lock-based critical sections in Real-Time Java [242]: methods are executed transactionally on a uniprocessor system, and a high-priority thread can preempt a low-priority thread and roll back the low-priority thread's tentative execution of a method.

Gottschlich and Connors also applied contention management to the context of real-time systems in which the CM should respect per-thread priorities [109]. Using commit-time validation, the CM provides an `abort_before_commit` hook, which is invoked when a transaction is about to commit successfully. This hook gives the transaction a chance to abort itself if it sees a possible conflict with a higher priority in-flight transaction: the transaction *could* commit, but *chooses not to* in order to avoid the higher priority transaction aborting. A second `permission_to_abort` hook is used with commit-time invalidation; in effect, the committing transaction must check that it is permitted to abort *all* of the victim transactions that it conflicts with. If it does not receive permission, then it aborts itself.

Gottschlich and Conners show how priority-inversion-style problems can occur with some TM design choices. For instance, if a transaction writing to a location cannot detect that concurrent transactions are reading from it, then a high priority read-only transaction may continually be aborted by low-priority writers. Spear *et al.* address this problem by making sure that readers are "visible" if they are operating at high priority [300].

2.3.3.2   Queuing and scheduling

Recent work has explored using forms of queuing to reduce contention. Intuitively, it should often be possible to predict if transactions TA and TB are likely to conflict and, if so, to then to run them sequentially rather than concurrently.

One option is for the programmer to provide information about possible conflicts. Bai *et al.* introduced a *key-based* technique for executing transactions that access dictionary-style data structures [26] in which transactions are partitioned between processors based on the keys that they are accessing. This can have two benefits: it prevents conflicting transactions from running in parallel, and it promotes locality since repeated accesses to the data representing the same key will be made from the same processor.

Yoo *et al.* describe a scheduling system based on an estimate of a process-wide conflict rate [340]. Under low contention, threads run transactions directly. If the conflict rate is high, then threads request that transactions are run via a centralized queue.

With CAR-STM, Dolev *et al.* maintain per-core queues of transactions that are ready to run [90]. Application threads place transactions on these queues, and dedicated transaction-queue threads (TQ threads) take transactions off the queues and execute them. If a transaction TA is aborted because of a conflict with transaction TB, then TA is moved to the queue that TB was fetched from. In effect, this serializes the transactions. As an extension, the *permanent serializing contention manager* keeps TA and TB together as a group if they conflict with a third transaction. This extension guarantees that any pair of dynamic transactions can conflict at most once. A *collision avoidance* mechanism attempts to avoid transactions conflicting even this much: an application can provide information about likely collisions, and CAR-STM uses this to decide where to place transactions.

Ansari *et al.* extend this work by running transactions on dedicated TQ threads, moving transactions between TQ threads according to the conflicts that occur, and also varying the *number* of TQ threads according to the abort rate: if aborts are frequent then the number of TQ threads is reduced to avoid running conflicting transactions concurrently, but if there are few aborts, then the number of TQ threads can safely be increased [18; 19; 20; 21].

Sonmez *et al.* investigated fine-grained switching between optimistic and pessimistic concurrency control, based on the level of contention that is seen; optimistic concurrency is used for data on which contention is low, and pessimistic concurrency is selected if contention is high. In practice, this approach is not always successful because, under the pessimistic implementation, the avoidance of wasted work is offset by an added risk of roll-backs to recover from deadlock [297].

Dragojević *et al.* investigated a *prediction-based* technique for deciding when to serialize transactions [94]. They predict the read/write-sets of a transaction based on the previous transactions executed by the same thread. If a high abort rate is seen, then transactions that are predicted to conflict are executed serially.

Kulkarni *et al.* show how application-specific knowledge is useful when scheduling work items in a system using optimistic concurrency control: heuristics that promote cache reuse for one workload can trigger pathological behavior in another (e.g., LIFO processing of newly generated work items) [181]. These observations may apply to TM, and similar controls may need to be exposed.

2.3.4   REDUCING CONFLICTS BETWEEN TRANSACTIONS

Some programming idioms simply do not fit well with TM. For instance, suppose that a program initially scales well, but the programmer wishes to add a statistics counter to track the number of times that a particular operation is used:

```
do {
    StartTx();
    int v = ReadTx(&count);
    WriteTx(&count, v+1);

    // Long-running computation
    ...
} while (!CommitTx())
```

Each thread's accesses to `count` will conflict. Worse, if lazy conflict detection is used, these conflicts may go undetected until *after* the long-running computation has been done. The conflicts on `count` can remove all of the possible concurrency between the transactions. Even worse, the programmer might not even care about detecting conflicts on `count` if erroneous values can be tolerated when gathering statistics.

Harris and Stipić described a taxonomy of "benign" conflicts where an implementation of TM would typically signal a conflict but where, at a higher level, the conflict does not matter to the application [139]:

- *Shared temporary variables.* This type of conflict occurs when transactions use shared data for temporary storage. The values do not need to be retained from one transaction to the next, but the accesses must typically be treated as conflicts. Fraser described this kind of conflict in a red-black tree implementation [105] in which transactions working near the leaves of the tree will write to a *sentinel node* whose contents do not need to be retained between transactions.

- *False sharing.* These conflicts occur when the granularity at which the TM detects conflicts is coarser than the granularity at which the application is attempting to access data. For instance, if a TM detects conflicts on complete cache lines, then transactions accessing different fields held on the same line will conflict. Lev and Herlihy introduced the terms *read-coverage* and *write-coverage* to distinguish these larger, coarser sets on which conflict detection is performed from the exact read-sets and write-sets that a transaction accesses [143].

- *Commutative operations with low-level conflicts.* This happens in our example: the complete increments to `count` are commutative, but the low-level reads and writes that they are built from will conflict.

- *Making arbitrary choices deterministically.* For instance, a `GetAny()` operation for an implementation of sets over linked lists may return the same item to two concurrent callers. However, using a non-deterministic implementation could reduce conflicts, by allowing concurrent transactions to remove different items.

These problems can be tackled at multiple levels in the system:

2.3.4.1 Value Prediction and Value-Based Validation

Tabba *et al.* show how speculation can be used in an HTM system to mitigate the effect of false conflicts caused by cache-line-granularity conflict detection [312]: a transaction can speculatively use stale cache lines, so long as the values that it used are checked when the transaction is subsequently validated. This means that a concurrent transaction can make updates to other parts of the same cache line without triggering an abort.

Similar techniques have been used in STM systems [77; 135; 236] (Section 4.4.2), and other systems based on speculative execution (e.g., [30; 88]).

2.3.4.2   Early Release and Elastic Transactions

Several early STM systems provided forms of *early release* operation [105; 146; 295; 298]. that allows a program to explicitly remove entries from a transaction's read-set or write-set. A recurring example is searching for an item in a sorted linked list; traversing the list will draw all of the items into the read-set, and so a search that finds an item near the end of the list will conflict with an insertion near the start. Early release can be used to discard the earlier items from the read-set since conflicts to them do not matter. Programming with early release can resemble hand-over-hand manual locking [149].

Using early release correctly requires great care. Correct usage depends on the complete set of operations that might manipulate the data structure. The "search" example is correct if the only other operations are simple insertions and deletions, but it is incorrect if a hypothetical "delete all greater than X" operation was added which truncates the list at X. That kind of operation would make it incorrect to use early release because the conflict with these concurrent bulk deletions could go undetected.

The second problem with early release is how it interacts with composition. Suppose that a list contains the elements 1..10 and a thread searches for 4 and then searches for 8. Early release during the search for 8 would remove the read-set entries for 4.

Elastic transactions allow a transaction to perform a prefix of operations which do not need to be serialized along with the "real" part of the transaction [104]. Conflicts can be tolerated during this prefix before maintaining a read-set and write-set as normal once the first update is made (or once a specific form of read operation is used). This provides cleaner support for composition than early release: atomicity extends from the first update to the end of the complete transaction even if multiple operations using elastic transactions are executed.

2.3.4.3   Tracking Dependencies

In some situations aborts can be avoided by explicitly tracking conflicts that occur between transactions, and avoiding aborting transactions so long as data is forwarded consistently between them— e.g., if transaction TA and transaction TB conflict, then this is acceptable so long as one of them always appears to run after the other. Circular dependence must either be prevented by stalling transactions, or the cycle must be avoided by aborting transactions. This observation can lead to an increase in concurrency between transactions (e.g., in Ramadan *et al.*'s HTM [254] and STM [255]). However, it does raise the specter of *cascading aborts* in which the abort of one transaction must trigger the aborts of any other transactions that have dependencies on it.

This kind of dependency-tracking could provide better scalability for our example using a shared counter: transactions using the counter can run concurrently, so long as their actual ReadTx/WriteTx pairs are not interleaved.

Pant *et al.* [239] and Titos *et al.* [317] examined the use of value-prediction hardware to provide a transaction with values that are expected to be written by a conflicting transaction. Pant *et al.* also examine a limited form of communication of uncommitted values [240], extending an HTM using eager version management with the ability for a transaction to read uncommitted state, with the reader performing additional checks to watch for the speculative state to be committed.

2.3.4.4 Partial Reexecution

An alternative to reducing the frequency of conflicts is to reduce the penalty that is incurred if they occur. *Intermediate checkpoints* can be used during a transaction's execution, representing a snapshot of the transaction's current state. If a conflict occurs after a checkpoint then execution can be restarted from the snapshot, rather than from the start of the whole transaction [175; 326].

Abstract nested transactions (ANTs) [139] provide a mechanism for partial re-execution of sections of a transaction. ANTs provide a hint to the TM implementation that the operations within the ANT should be managed separately from the rest of the enclosing transaction: for instance, an STM-based implementation may keep separate logs for an ANT so that it can be reexecuted without rerunning the complete transaction. ANTs would work well for our example of a shared counter: the ReadTx/WriteTx pairs would be placed in an ANT, and these sub-transactions could be re-executed if conflicts occur. Currently, ANTs have been implemented only in STM-Haskell.

2.3.5   HIGHER-LEVEL CONFLICT DETECTION

Contention management (Section 2.3.3) and conflict-reduction techniques (Section 2.3.4) all provide ways of reducing the amount of contention experienced as transactions execute. However, they are fundamentally limited by the fact that transactions are expressed using ReadTx/WriteTx operations. As with our simple example using a shared counter, a transaction can often be using abstractions that commute at a higher level but which conflict in terms of the actual reads and writes that they perform. Chung *et al.* give a practical example of this kind of problem in a transactional version of the SPECjbb2000 workload [63].

This observation has led to several techniques that allow the programmer to raise the level of abstraction at which a TM system operates, introducing higher-level notions of conflict detection: e.g., if the TM system "knows" that a given ReadTx/WriteTx pair is performing an increment, then it can permit concurrent transactions to execute while incrementing the same shared counter.

All of the systems for performing higher-level conflict detection rely on a notion of explicitly-defined abstract data types, which are required to completely encapsulate the shared state that they are managing (Figure 2.6). This requires, for instance, that *all* of the accesses to the shared counter should be made via Inc/Dec/Read operations that the TM system manages, rather than being made by direct access to the underlying shared memory. Agrawal *et al.* developed a type system to enforce this kind of discipline [14].

For each abstract data type (ADT), the programmer must specify the following:

- *Commutativity.* In terms of their effect on the abstract state, which pairs of operations commute with one another? For example, Inc and Dec commute with one another, but they do not commute with Read.

- *Inverses.* If a transaction rolls back after performing an operation on the ADT, then how should its work on the ADT be undone? In this case, an increment by x should be compensated by

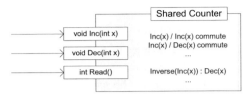

Figure 2.6: An abstract data type (ADT) implementing a shared counter. The counter's value is accessed only via the Inc, Dec, and Read operations. The ADT defines which operations commute with one another and what their inverses are.

a decrement by x. On a shared counter, Read is a *disposable* operation which does not require an inverse.

This leads to a need for concurrency control at two levels. First, there must be *high-level concurrency control*, to make sure that the operations executed by two concurrent transactions commute with one another. Typically, this is implemented by pessimistic concurrency control using *abstract locks*, which are acquired by a transaction's operation on an ADT, and held until the transaction commits. For a shared counter, there would just be one abstract lock, acquired in one mode for Inc and Dec, and acquired in the other mode for Read. For a shared set, there could be a separate abstract lock for each key in the set.

In addition to this high-level abstract locking, there must be *low-level concurrency control* within the implementation of the ADT. For instance, even though two transactions may execute Inc, the individual increments must be performed atomically. One approach is to use *open nested transactions* (Section 2.2.6) to do this: the Inc operation itself runs inside a transaction but commits immediately to memory. This serializes it with respect to other operations on the ADT.

An alternative approach to using open nesting is to rely on an existing thread-safe data structure's implementation to provide this low-level concurrency control. Herlihy and Koskinen introduced this technique as *transactional boosting* [142; 177]. Unlike open nesting, transactional boosting allows the existing thread-safe data structure's implementation to execute *non transactionally*, using its own internal synchronization, rather than assuming that the data structure's implementation should be executed within a sub-transaction.

More generally, in addition to commutativity, we can exploit knowledge of how different operations can be reordered with respect to one another in controlled ways [178]—e.g., using a model like Lipton's notions of left-movers and right-movers [195]. Similar ideas of high-level conflict detection have been used in other transactional systems [329] and in the Galois system for optimistic parallelization [181; 182].

2.4 SUMMARY

In this chapter, we have introduced a basic programming interface for TM, sketched the main design choices that are taken in TM implementations, and examined the use of TM from the point of view of its semantics and the kinds of progress properties that can be provided.

To conclude this chapter, we return to a series of trade offs that have arisen at several points.

First, there is the tradeoff between the semantics provided by a TM interface and the range of implementation techniques that are available. This was not always understood in earlier work where an "improvement" to a TM implementation might reduce the set of programs that it supports. It is, therefore, important, when evaluating different practical techniques, to identify whether or not they support the same set of programs. The classification of programming disciplines such as static separation and TDRF provides one way of doing this, with reference to a common definition of TSC rather than the details of a given implementation.

Second, there is the tradeoff involving concurrency between transactions and the performance of a TM implementation. Many papers have shown, informally, that greater concurrency can be supported by keeping additional information about how transactions have interacted—but the amount of bookkeeping needed to identify these opportunities has often been non-trivial, and if a workload has a low level of contention, then it has little opportunity to benefit from any additional concurrency.

Finally, when looking at the practical performance of TM systems, many of the existing implementations involve both a careful design of algorithms and care over the low-level details of the code; e.g., how data is laid out in memory in software or exactly which messages might be added to a directory protocol in hardware. Separating out the degree to which "good algorithms" versus "good code" contribute to performance is difficult, and firm results favoring one kind of technique over another are rare.

CHAPTER 3

Building on Basic Transactions

The TM programming interface from Chapter 2 can be very cumbersome to use: reads and writes must be written verbosely as `ReadTx/WriteTx` calls, and the logic of an algorithm is obscured by all the boilerplate involved in starting and committing transactions. The basic TM interface has other problems too, beyond its verbosity. The exact interface is often rather coupled to a particular TM implementation; e.g., whether or not an explicit `AbortTx` is available and whether or not reexecution of failed transactions can occur at any time. This limits portability of programs using TM. The basic interface does not provide any constructs for *condition synchronization* where one thread wishes to wait until a data structure is in a given state (e.g., a queue is not full): some form of integration with condition synchronization is necessary if TM is to be used as a general-purpose abstraction for concurrency control. The basic interface is not good from a programmability viewpoint because it requires code to be written explicitly to use TM, making it hard to re-use common libraries.

In this chapter, we look at how to extend the simple TM interface to remedy some of its shortcomings and how to build higher-level language constructs that are portable across a range of implementations. First, we look at `atomic` blocks as a mechanism to delimit code that should execute atomically and in isolation from other threads (Section 3.1); `atomic` blocks avoid the need to call explicit `ReadTx/WriteTx` operations by hand. We examine the design and implementation of `atomic` blocks and how their semantics relate to those of TM.

We then look at how to extend `atomic` blocks with support for condition synchronization, and integration with non-TM features like garbage collection (GC), system calls, and IO (Section 3.2).

Section 3.3 looks at current experience programming with TM—tools such as debuggers and profilers, workloads based on TM or `atomic` blocks, and user studies about how these techniques compare with lock-based programming.

Finally, in Section 3.4, we survey research on alternatives to `atomic` blocks. This includes work on "transactions everywhere" models, in which transactional execution in the norm, and models in which TM is used as an implementation technique for ordinary critical sections.

3.1 BASIC ATOMIC BLOCKS

Since early work on language constructs for TM, a prevalent idea has been to provide some kind of construct for marking an `atomic` block of statements. The implementation wraps this block in operations to manage a transaction and replaces the memory accesses that occur within the block with `ReadTx/WriteTx` calls. Implementations of `atomic` blocks have been developed for many languages, including C/C++ [232; 327; 336], C# [138], Haskell [136], Java [134; 220; 336; 345], OCaml [266], Scheme [170], and Smalltalk [259].

Various keywords are used in the literature, but for uniformity in this book, we write examples using `atomic` blocks in a Java or C#-like language [50; 134; 136]. Unsafe languages, such as C/C++, may require different implementations or place different requirements on the kinds of TM that they can use, but the language features are likely to be similar. There is substantial ongoing work examining these topics [8; 73; 232; 287].

As in the previous chapter, terminology differs somewhat among different viewpoints for these constructs. Referring to the ACID properties of transactions, `atomic` is a misnomer or at least an over-simplification because TM provides *isolation* as well as atomicity. An alternative viewpoint is that "atomic" should be taken to mean "indivisible" rather than the "all or nothing" of ACID. From this viewpoint, an `atomic` block is one whose effect on the process's state is that of an indivisible update.

Many of the concepts for `atomic` blocks were anticipated in a paper by Lomet in 1977 [199], which was published soon after the classic paper by Eswaran [101] on two-phase locking and transactions. Lomet reviewed the disadvantages and shortcomings of synchronization mechanisms, such as semaphores, critical regions, and monitors, and noted that programmers used these constructs to execute pieces of code atomically. His suggestion was to express the desired end directly and shift the burden of ensuring atomicity onto the system. An atomic action appears atomic from a programmer's perspective, as no other thread or process can see the intermediate steps of an action until the action completes and commits.

Returning to `atomic` blocks and the dequeue example from Chapter 2, the `PushLeft` method could be written as follows:

```
void PushLeft(DQueue *q, int val) {
  QNode *qn = malloc(sizeof(QNode));
  qn->val = val;
  atomic {
    QNode *leftSentinel = q->left;
    QNode *oldLeftNode = leftSentinel->right;
    qn->left = leftSentinel;
    qn->right = oldLeftNode;
    leftSentinel->right = qn;
    oldLeftNode->left = qn;
  }
}
```

The semantics of the construct are subtle, but for now, assume that the block executes with the failure atomicity and isolation properties of a transaction, and that the contents of the block are automatically reexecuted upon conflict so that it appears to run exactly once. The resulting transaction is dynamically scoped—it encompasses all code executed while control is in the atomic block, regardless of whether the code itself is lexically enclosed by the block—transactional execution extends into any functions called from within the block.

Some languages also define atomic functions or methods, whose body executes in an implicit atomic statement [242]:

```
void Foo() {                             atomic void Foo {
    atomic {                                 if (x != null) x.foo();
        if (x != null) x.foo();              y = true;
        y = true;                        }
    }
}
```

$\Longleftrightarrow$

Composability

A key advantage of `atomic` blocks over lock-based critical sections is that the `atomic` block does not need to name the shared resources that it intends to access or synchronize with; it synchronizes implicitly with any other `atomic` blocks that touch the same data. This feature distinguishes it from earlier programming constructs, such as monitors [156] or conditional critical regions [157], in which a programmer explicitly names the data protected by a critical section. It also distinguishes `atomic` blocks from constructs like Java's `synchronized` blocks, or C#'s `lock` regions.

The benefit of this is a form of *composability*: a series of individual atomic operations can be combined together, and the result will still be atomic. This does not hold with lock-based abstractions. Consider a list that uses internal synchronization to provide linearizable `Insert` and `Remove` operations on a set of keys. With locks, there is no immediate way to combine these operations to form a new `Move` operation that deletes an item from one list and inserts it into another—either one must break the encapsulation of the list and expose whatever kind of concurrency control it uses internally, or one must wrap additional concurrency control around all of the operations on the list. Conversely, if the list were implemented with `atomic` blocks, one could write:

```
bool Move(List s, List d, int item) {
    atomic {
        s.Remove(item); // R1
        d.Insert(item); // I1
    }
}
```

As a result, an `atomic` block enables abstractions to hide their implementation and be composable with respect to these properties. Programming is far more difficult and error-prone if a programmer cannot depend on the specified interface of an object and instead must understand its implementation. Using locks in application code to achieve mutual exclusion exposes this low level of detail. If a library routine accesses a data structure protected by locks A and B, then all code that calls this routine must be cognizant of these locks, to avoid running concurrently with a thread that might acquire the locks in the opposite order. If a new version of the library also acquires lock C, this change may ripple throughout the entire program that uses the library.

An `atomic` block achieves the same end but hides the mechanism. The library routine can safely access the data structure in a transaction without concern about how the transactional properties are maintained. Code calling the library need not know about the transaction or its implementation nor be concerned about isolation from concurrently executing threads. The routine is composable.

Limitations of Atomic Blocks

The `atomic` block is not a parallel programming panacea. It is still, regrettably, easy to write incorrect code, and writing concurrent code remains more difficult than writing sequential code. However, there are two additional problems when programming with `atomic` blocks:

Many prototype implementations lack *orthogonality* between `atomic` blocks and other language features: either there are some language features that cannot be used in `atomic` blocks, or there are some settings where `atomic` blocks cannot be used. Operations with external side-effects (e.g., IO) are the most prominent example of non-orthogonality; we look at many problems of this kind in Section 3.2 along with the techniques developed to support them.

The second problem is that, even with orthogonality, some programming idioms do not work when they form part of an atomic action. A particular problem occurs with barrier-style operations which involve synchronization between threads. Here is an example from Blundell *et al.* [35; 36], which illustrates the kind of problem that occurs when enforcing atomicity around an existing synchronization construct:

```
volatile bool flagA = false;
volatile bool flagB = false;

// Thread 1                    // Thread 2
atomic {                       atomic {
  while (!flagA); // 1.1         flagA = true;
  flagB = true;   // 1.2         while (!flagB);
}                              }
```

The code in the `atomic` blocks is incorrect since the statement 1.1 can only execute *after* Thread 2's `atomic` block, but the statement 1.2 can only execute *before* Thread 2's `atomic` block: hence it is not possible for these two blocks both to run atomically and with isolation. Although this example is contrived, it illustrates a kind of problem that can occur if the synchronization is buried within a library that the two `atomic` blocks use: either the use of the library needs to be prohibited inside `atomic` blocks, or the library needs to be written to use some other synchronization mechanism instead of transactions. We return to this kind of example when looking at IO in Section 3.2.

3.1.1 SEMANTICS OF BASIC ATOMIC BLOCKS.

Early work that provided `atomic` blocks often defined their semantics implicitly, based on the properties of a particular implementation. As with examples such as "privatization" in the previous chapter, code that one might expect to work often failed on early implementations, and code that worked on one variant of `atomic` blocks did not necessarily work on others. One might prefer a language construct to provide clearly-defined semantics, and to permit a range of different implementations (much like there are many GC algorithms, suited to a variety of workloads, but providing the same abstraction to a programmer [114]).

It is useful to distinguish the semantics of `atomic` blocks from those of TM. There are several reasons: we may wish to support implementations of `atomic` blocks that do not use TM at all (e.g., using lock-inference techniques [59; 74; 129; 151; 214]). Second, recent implementations of `atomic` blocks are complex and may involve work in many parts of a language implementation in addition to the use of TM: e.g. transforming a program to sandbox zombie transactions (Section 3.1.3), or transforming a program to improve performance (Section 4.2.2). It, therefore, makes sense to define the public interface of `atomic` blocks seen by the programmer, in addition, to any private interfaces used within a given implementation.

Many possible semantics for `atomic` blocks are discussed in the literature. In this book, we focus on two specific designs, mirroring the approaches taken in Chapter 2:

3.1.1.1   Single-Lock Atomicity (SLA) for Atomic Blocks

As with TM, a notion of SLA is appealing because many programming languages already include locking. We can apply the idea of SLA to the definition of `atomic` blocks, much as we did to the definition of TM: each `atomic` block behaves as if it executes while holding a single, process-wide lock. This lock-based version then serves as a reference implementation for the behavior of the program using `atomic` blocks. The question of how a given program should behave using `atomic` blocks is then reduced to the question of how the lock-based version behaves under the language's existing semantics. This provides a methodical approach for defining how other language constructs interact with `atomic` blocks: they should behave the same as they would with a single lock.

Modern languages typically define which lock-based programs are considered race-free [43; 207]. Variants of SLA can either require that an implementation is faithful to the lock-based reference for *all* programs—or, more weakly, require that the implementation is faithful only when the lock-based program is race-free. Menon *et al.* illustrate how supporting *all* programs consistently with the Java memory model can complicate the implementation of SLA or undermine its performance [219; 220].

Using locks to define the semantics of `atomic` blocks raises the same kinds of problems that can occur when using locks to define the semantics of TM—e.g., it is not clear how to extend the definition to include constructs such as open nesting, since there is no clear lock-based counterpart for these.

However, there is an additional difficulty when using lock-based reference models with `atomic` blocks. When defining the semantics of TM, weaker notions such as disjoint lock atomicity (DLA), asymmetric lock atomicity (ALA), and encounter-time lock atomicity (ELA) provided additional models that allowed faster implementations than SLA. These designs are problematic when used to define semantics for `atomic` blocks. This is because DLA, ALA, and ELA are defined in terms of the data accesses made by a transaction: these accesses are explicit with a `ReadTx/WriteTx` interface, but they are implicit when using `atomic` blocks, and the accesses made may usually be changed during optimization. For instance, consider this variant on the "empty publication" example under DLA:

```
// Thread 1                    // Thread 2
data = 1;                      atomic {
atomic {                         int tmp = data;
  int r = flag;                  flag = 42;
}                                if (ready) {
ready = 1;                         // Use tmp
                               } }
```

Like the original "empty publication" example, this involves a race on data. However, under SLA, the Java memory model nevertheless guarantees that if Thread 2 sees ready==1 then it will also see data==1. In addition, under DLA, the programmer may hope that the synchronization on flag means that the two atomic blocks must synchronize on a common lock, and so Thread 2 will still see the publication from Thread 1. However, it seems equally reasonable for an optimizing compiler to remove Thread 1's read from flag because the value seen is never used. With this read removed, there is no synchronization between the threads under DLA, and so the publication is not guaranteed.

The crux of the problem is that DLA, ALA, and ELA can be used in defining the semantics of programs using a TM interface explicitly, but they are less appropriate when defining the semantics of atomic blocks because ordinary program transformations can remove memory accesses and, therefore, change the sets of locks involved. We, therefore, stick with SLA as one model for the semantics of atomic blocks.

3.1.1.2 Transactional Sequential Consistency (TSC) for Atomic Blocks

In Section 2.2.5, we discussed the use of transactional sequential consistency (TSC) as an mechanism for defining the semantics of transactions in a way that, unlike SLA, avoided referring to existing synchronization constructs.

With transactions, TSC requires that the effects of a transaction being attempted by one thread were not interleaved with any operations being executed by other threads. The same idea can be applied at the level of atomic blocks: under TSC, if an atomic block is being executed by one thread, then no other operations can appear to be interleaved from other threads. Once again, the "appear to" is important: an *implementation* of TSC may execute threads in parallel, and they perform all kinds of low level interleaving between their operations. However, if the implementation is to provide TSC, then the effects of these low-level interleaving must be invisible to the programmer.

Figure 3.1 contrasts SLA with TSC: both models allow the executions shown in Figure 3.1(a)–(b), but TSC prohibits the execution in Figure 3.1(c) in which the work being done outside the atomic block is interleaved with Thread 1's execution of the atomic block. For simple workloads, where atomic blocks just contain accesses to shared memory, case (c) can only occur if the program has a data race—in which case the distinction can appear academic if we are only interested in race-free programs.

(a) Execution of Thread 2's non-transactional work before Thread 1's `atomic` block (valid under SLA and TSC).

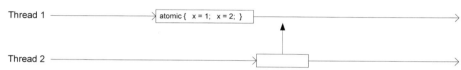

(b) Execution of Thread 2's non-transactional work after Thread 1's `atomic` block (valid under SLA and TSC).

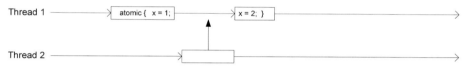

(c) Execution of Thread 2's non-transactional work during Thread 1's `atomic` block (only valid under SLA, not valid under TSC).

Figure 3.1: Contrasting SLA with TSC. In this example, Thread 1 executes an `atomic` block, and Thread 2 executes other work, without using an `atomic` block.

However, the differences between SLA and TSC become more significant when we consider how `atomic` blocks might interact with other language features, in addition to data in shared memory. For example, suppose that a program executes the following two pieces of work concurrently:

```
// Thread 1                    // Thread 2
atomic {
  print("Hello ");
                               print("brave new");
  print("world!");
}
```

The output "Hello brave new world!" is valid under SLA: Thread 2 does not require the lock held by Thread 1's `atomic` block. However, under TSC, this output is prohibited because TSC requires that no other thread's work is interleaved within Thread 1's `atomic` block. We return to many examples of this kind throughout the chapter to illustrate the subtleties that exist in defining the semantics of `atomic` blocks, and the differences between options that have been explored in the literature.

The use of TSC to define the semantics of atomic blocks builds on definitions such as the operational semantics for STM-Haskell [136], or the high-level semantics of Abadi *et al.* [3] and Moore and Grossman [226]. In each case, these semantics define a simple time-sliced model of execution, in which threads execute complete atomic blocks without any other thread's operations being interleaved. This kind of definition has also been termed a *strong semantics* [3].

3.1.2   BUILDING BASIC ATOMIC BLOCKS OVER TM

With SLA and TSC in mind as definitions for the semantics of atomic blocks, we now turn to the main implementation techniques used in building atomic blocks over TM. A basic implementation starts by expanding the block itself into a loop that attempts to execute the contents in a series of transactions:

```
atomic {                    do {
    ...          ⟹            StartTx();
}                               ...
                            } while (!CommitTx());
```

Intuitively, the loop will continue attempting the body of the atomic block until it manages to commit successfully. The body of the loop must be compiled to use TM for its memory accesses. With the example API from Chapter 2, this means replacing each read and write with a call to the corresponding ReadTx/WriteTx operation.

We must also ensure that any function calls or method calls made from within the atomic block will also use the TM API. Some implementations use dynamic binary translation to do this [236; 328; 339], but a more common technique is to generate specialized versions of functions during compilation. For instance, in the Bartok-STM implementation of atomic blocks for C#, each method Foo logically has a TM-aware version named Foo$atomic [138]. Each method call within an atomic block is replaced by a call to the corresponding $atomic method. Similarly, within an $atomic method, memory accesses must be expanded to use ReadTx/WriteTx and calls modified to use $atomic methods. The Deuce-STM system performs this kind of rewriting dynamically during class loading in a Java Virtual Machine [174].

We call this *explicit instrumentation*. Implementing explicit instrumentation can require particular care in the handling of indirect branches—e.g., calls to virtual methods or calls via function pointers. The challenge is in making sure that an instrumented version of the target code is available, and that this is correctly executed when performing a call inside an atomic block. Harris and Fraser developed techniques to handle virtual function calls by maintaining a second virtual-method-table per class [134]. Wang *et al.* discuss techniques to handle function pointers [327] by using static analyses to identify the functions that might be possible targets. STM systems based on dynamic binary translation [236; 328; 339] handle both of these cases by simply translating the target of an indirect branch if necessary. Detlefs and Zhang developed techniques to instrument accesses to by-ref parameters passed in to functions containing atomic blocks [81].

Some TM implementation *implicitly* perform transactional reads and writes whenever a transaction is active—this is particularly true of HTM systems. In these, it is unnecessary to generate

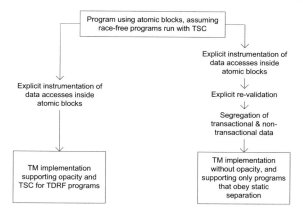

Figure 3.2: Two example ways to support TSC for race-free programs over different underlying TMs: strong guarantees could be inherited from the TM itself, or they could be built over a weaker TM.

separate `$atomic` versions of methods for data accesses, although there can be some cases where other kinds of specialization are useful—e.g., if a TM-aware memory allocator is to be used (Section 3.2.5), or if performance could be improved by distinguishing accesses to thread-local data.

Function Compatibility.   Many implementations of `atomic` blocks provide facilities to indicate which methods can be used inside them and which cannot. This is particularly useful for library methods whose behavior is incompatible with atomicity—e.g., the barrier example from Section 3.1 should not be used inside `atomic` blocks, and so it may be better to discover this kind of use at compile-time.

The STM.NET system provides an `AtomicNotSupported` attribute that can be used to label this kind of function, along with related attributes to indicate functions which can be used only inside `atomic` blocks. As an optimization, annotations have also been used to identify functions that can be used safely inside an `atomic` block without any instrumentation—Ni *et al.* call this a `tm_pure` function [232].

3.1.3   PROVIDING STRONG GUARANTEES OVER WEAK TM SYSTEMS

How should an implementation of `atomic` blocks provide semantics such as SLA or TSC? In both of these semantics, idioms like the privatization and publication examples from the previous chapter must be supported, and implementation details like data-access granularity must be hidden.

Figure 3.2 sketches two of the possible alternatives. The first, on the left, is to provide strong guarantees in the TM itself: for instance, ensuring that there are no granular lost update (GLU) problems caused by the TM's writes spilling over onto adjacent locations, providing opacity so that transactions see a consistent view of memory as they run, and either providing strong isolation, or, at least, sufficient constraints on transaction ordering that transactional-data-race-free (TDRF) programming idioms work. This is straightforward and natural if the underlying TM provides strong guarantees—e.g., if the `atomic` block can be supported over an HTM system with strong isolation.

However, many implementation techniques have been developed that allow strong language-level properties to be provided without necessarily providing them at the TM level. This is the right-hand path on Figure 3.2. In effect, this provides an additional layer in the language run-time system, above the TM, to ensure that the remainder of the system is shielded from whatever anomalous behavior might be produced. As a running example of this second approach, consider an implementation that is building on a TM which does not provide opacity and which supports only static separation (i.e., where any given piece of shared mutable data must be accessed either always transactionally or always non-transactionally). Other options exist, but we focus on this one because it provides a challenging setting: a naïve implementation will exhibit a whole range of anomalies.

3.1.3.1   Detecting Invalid Transactions

When building on a TM without opacity, it is necessary to detect if a transaction becomes invalid, and to sandbox its effects in the meantime. For instance, consider a simple example with a transaction that enters a loop depending on the values it sees in x and y:

```
atomic {
  if (x != y) {
    while (true) {
      /*W1*/
    }
  }
}
```

If the program's `atomic` blocks maintain an invariant that x==y, then this example could enter an endless loop at W1 if it sees inconsistent values in the two variables. This example is contrived, but more complex practical examples have been observed—e.g., Fraser reports a problem with a transaction looping while traversing a particular kind of red-black tree implementation [105].

To detect this kind of invalidity, the program must usually be compiled to include explicit validation checks. Standard analyses can be used to optimize the placement of these validation checks, while still ensuring that they are executed on all looping code paths. The implementation of this additional validation work can also be guarded by a thread-local test so that "real" validation is only done, say, every 1M iterations; this adds little delay to detecting invalidity, but it reduces the overhead that checking imposes. Alternatively, if a periodic timer is available, then it can be used to set a validation-request flag or to directly perform validation itself.

3.1.3.2   Sandboxing Invalidity and Building Strong Isolation

In addition to detecting invalidity, our example implementation must sandbox the effects of an invalid transaction—for instance, instead of entering a loop at W1, the program might attempt to write to arbitrary memory locations. One way to do this is to build strong isolation over the weaker guarantees that a given TM provides.

Hindman and Grossman [154; 155] and Shpeisman *et al.* [288] introduced early implementations of strong isolation in this kind of setting. The basic idea is to expand non-transactional accesses into optimized short transactions so that they detect whether or not there are conflicts. This means that conflicts between transactional and non-transactional accesses will be handled correctly by the TM.

When compared with full transactions, these optimized versions can be specialized to perform a single memory access, rather than needing to maintain logs in the normal way. Sequences of related accesses (e.g., read-modify-write to a single location) can also be aggregated and specialized. In addition, specialization may be possible within the implementation according to the particular TM being used—e.g., whether or not eager version management is used.

Static analyses can also be used to improve the performance of this approach by identifying non-transactional accesses that need not communicate with the TM—e.g., because the data involved is immutable, thread-local, or never accessed transactionally. In each of these cases, there is no possible conflict between the transactional and non-transactional accesses. However, note that analyses must consider the accesses made by the *implementation* of transactions, rather than just the accesses that the programmer might anticipate from the source code under TSC or SLA—for instance, an analysis must consider the granularity of accesses to avoid GLU problems and the effects of zombie transactions.

Shpeisman *et al.* introduced a "not accessed in transactions" (NAIT) analysis that identifies which non-transactional accesses cannot conflict with transactional accesses. Schneider *et al.* investigated enhancements of this approach in a JIT-based implementation [281], using a dynamic variant of NAIT (D-NAIT) and additional escape analyses. They exploit the fact that all of the heap accesses are made by code generated by the JIT: the results of the NAIT analysis therefore only need to be sound for this body of actually-runnable code, rather than for the larger set of code that makes up the application and the libraries it may use. This observation reduces the number of memory accesses that need to be expanded to interact with the TM. Indeed, it can mean that a program does not need to have any accesses expanded until it starts to use transactions, enabling a "pay to use" implementation. In addition, a JIT-based approach can accommodate dynamically-generated byte-code functions. Bronson *et al.* investigated feedback-directed selection between different customized barrier implementations in an implementation of strong isolation [44].

An alternative technique is to use memory-protection hardware to isolate data being used inside `atomic` blocks from data that is being used directly. Baugh *et al.* showed how this can be done using hardware extensions for fine-grained user-level protection [28] (Section 5.5.4). Conventional page-based memory protection hardware can also be used, albeit in a rather complex design [5]. In this latter approach, the process's virtual address space is organized so that there are two *views* of the same heap data mapped at different virtual address ranges. One view is used for transactional accesses, and the other view is used for non-transactional accesses. The TM implementation revokes access to the normal view of a page before accessing it within a transaction. Thus, conflicting non-transactional accesses will incur page faults. To avoid disastrous performance, changes to page permissions can be made lazily (rather than switching them back-and-forth as each transaction starts and commits), dynamic code updating can be used to replace instructions that generate frequent access violations by short transactions, transactional and non-transactional allocations can be made on separate pages (to discourage false sharing), and NAIT-style static analyses can be used to identify operations that cannot produce transactional/non-transactional conflicts.

3.1.3.3 Summary

As we have shown in this section, there are several techniques that can be put together to build SLA or TSC over a weaker TM. Whether or not these are needed in a particular implementation depends both on the language guarantees to be offered, and, of course, on the underlying properties of the TM.

For a specific target semantics, such as supporting race-free programs with SLA or TSC, it is not yet clear exactly which combination of these approaches works best. For instance, techniques to build strong isolation can be very complicated to implement and introduce additional instrumentation at run-time. On the other hand, by building on a TM with weaker guarantees, a wider range of TM systems can be used, and it is possible that these may be faster than TMs which natively provide stronger guarantees.

3.2   EXTENDING BASIC ATOMIC BLOCKS

Basic atomic sections built over TM lack many of the features that programmers expect. There is no way to express *condition synchronization* where one thread's `atomic` block needs to wait for another thread. This severely limits the applicability of `atomic` blocks—for instance, it prevents a blocking producer-consumer queue from being written. In addition, as we said at the start of the chapter, basic `atomic` blocks do not provide orthogonality with other language features—the simple implementation that we sketched does not provide mechanisms for performing IO within an `atomic` block or for using existing programming abstractions that might be implemented using locks. It is also unclear how language runtime services interact with the implementation of `atomic` blocks—e.g., to handle memory allocations or to operate the garbage collector.

We examine all of these questions in this section. First, we consider techniques to extend `atomic` blocks with support for condition synchronization (Section 3.2.1). Then, we consider support for *failure atomicity* and how this can be provided by integration with a language's exception mechanisms (Section 3.2.2). Finally, we look at integration between `atomic` blocks and a series of non-TM facilities such as IO and system calls (Section 3.2.3–3.2.7).

3.2.1   CONDITION SYNCHRONIZATION

In database systems, transactions are generally independent operations on a database. However, within a single process, it is often necessary to coordinate which operations execute and when—for instance, one `atomic` block stores an item into a shared data structure, and another block wants to wait until the item is available.

3.2.1.1 Retry and OrElse

Harris *et al.* [136] introduced the `retry` statement to coordinate `atomic` blocks in STM-Haskell, Ringenburg and Grossman introduced an analogous statement for a variant of ML [266], and Adl-Tabatabai *et al.* developed implementations for the Java programming language [7]. Semantically, an `atomic` block is permitted to run only when it can finish without executing any `retry` statements; `retry` can be seen as indicating that the `atomic` block has encountered a state in which it is not yet ready to run. For instance, one could write:

```
public void TakeItem() {
   atomic {
      if (buffer.isEmpty()) retry;
      Object x = buffer.GetElement();
      ...
   }
}
```

If the buffer is empty then the `retry` statement indicates that the `atomic` block is not yet ready to execute. This example illustrates the advantage of combining `atomic` and `retry` statements: since the predicate examining the buffer executes in the `atomic` block, if it finds an item in the buffer, then this value will be present when the next statement in the block goes to get it. The transaction's isolation ensures that there is no "window of vulnerability" through which another thread can come in and remove an item. Unlike explicit signaling, with conditional variables and `signal`/`wait` operations, `retry` does not need to name either the transaction being coordinated with, or the shared locations involved.

Operationally, when building `atomic` blocks over TM, a transaction that executes a `retry` statement simply aborts and then re-executes at some later time. Typical implementations of `retry` work by delaying this re-execution until another `atomic` block has updated one of the locations involved—in this example, that might mean waiting until a flag in the buffer is updated, but in more complex cases, a series of locations may need to be watched.

The `retry` operation provides a form of composable blocking, in the sense that a thread may group together a series of atomic operations into a larger compound atomic operation. For instance, to take two successive items from a buffer:

```
public void TakeTwo() {
   atomic {
      buffer.TakeItem();
      buffer.TakeItem();
   }
}
```

This example also illustrates the interaction between `retry` and nesting—semantically, a `retry` in either `TakeItem` operation signals that the entire nest is not yet ready to run. When built over TM, this can be implemented by rolling back to the very start of the outermost atomic block. If lazy version management is used, then optimized implementations could defer complete roll-back until it is known exactly which data is written by a conflicting transaction.

Harris *et al.* [136] also introduced the orElse operation: if X and Y are transactions then X orElse Y starts by executing X:

1. If X commits, the orElse statement terminates without executing Y.

2. If X executes a retry statement, the tentative work done by X is discarded, and the orElse operation starts executing Y instead.

3. If Y commits, the orElse statement finishes execution.

4. If Y executes a retry statement, the orElse statement as a whole executes retry.

The orElse statement must execute in an atomic block because the composition of two transactions is itself a transaction that can commit, abort, or retry.

For example, suppose we want to read a value from one of two transactional queues, and that GetElement is a blocking operation that will retry if a queue is empty. The following code checks the queues and returns a value from Q1 if possible; otherwise, it attempts to dequeue an item from queue Q2. As a whole, the code only retries if both are empty:

```
atomic {
    { x = Q1.GetElement(); }
    orElse
    { x = Q2.GetElement(); }
}
```

Harris *et al.* deliberately specified left-to-right evaluation of the alternatives, as opposed to running them concurrently or in non-deterministic order. The motivation for left-to-right evaluation is that it enables orElse to be used to adapt blocking operations into ones that produce a failure code. For instance, the following code returns null if a blocking GetElement operation tries to wait on an empty queue:

```
Object GetElementOrNull () {
    atomic {
        { return this.GetElement(); }
        orElse
        { return null; }
    }
}
```

Some practical systems provide a mechanism to limit the number of reexecutions [50]—variants of retry could include a maximum re-execution count, a time-out for waiting, or a time-out for the block's complete execution.

The semantics of `retry` and `orElse` enable many examples to be written concisely and, as shown above, they provide forms of composition that are not present when using locks and condition variables. However, a disadvantage of `retry` and `orElse` is that their behavior needs to be explained either in terms of prescience or in terms of speculation and aborts—both of these notions can be unintuitive for some programmers and difficult to implement when building `atomic` blocks over lock inference rather than over TM.

3.2.1.2 Explicit Watch Sets

There are several alternatives to `retry`, and `orElse`. Some systems provide a form of `retry`, which specifies an exact *watch set* [50; 266]. Only variables in the watch set need to be monitored while waiting. Consequently, a watch set can be smaller than the transaction's complete read set. Reducing the size of this set can improve performance, particularly for hardware, which may only be able to watch for updates on a limited number of locations.

The use of explicit watch sets fits well with the *dual data structure* style of concurrent data structures pioneered by Scherer and Scott [279]. In a dual data structure, blocking operations are expressed by adding explicit *reservations* to the structure, and arranging for updates that satisfy a reservation to wake the reserver—e.g., a dual stack would hold records representing pop operations that are attempted when the stack is empty. In this framework, `atomic` blocks can be used to update the data structure to add reservations, and a single-entry watch set can be used to monitor a flag within the reservation.

However, explicit watch sets do not provide the same kind of composability as `retry`, and programming with them requires care to ensure that all of the necessary locations have been considered. For instance, it might be straightforward to write something like "`retry full`" to watch the `full` flag of a single buffer, but this would be insufficient in a more complex example such as:

```
atomic {
   if (!stop) {
     buffer.TakeItem();
   }
}
```

In this case, the thread should watch the `stop` flag as well as `full`, because the decision to access the buffer is dependent on the value of `stop`. To avoid this kind of problem, one could restrict the use of watch sets to simple non-nesting `atomic` blocks, and one could require the `retry` operation to be lexically inside the block (as with some forms of roll-back [8]).

3.2.1.3 Conditional Critical Regions

Harris and Fraser proposed a form of guard [134], using a syntax modeled after conditional critical regions. For example, to decrement x when its value is positive, a programmer could write:

```
atomic (x > 0) {
   x--;
}
```

This syntax makes the guard condition explicit at the start of the `atomic` block, and so it might encourage programmers to do as little computation as possible before deciding whether or not to continue executing. Consequently, this may lead to smaller watch sets than with a `retry` operation. However, the semantics of the original implementation are essentially the same as `retry` because the condition could be an arbitrary expression, including a call to a function with side-effects. The side-effects would be rolled back if the condition was false.

Note that if a thread is waiting on entry to a conditional critical region in a nested `atomic` block, then it must still watch the complete read-sets for the enclosing blocks: as with our example with the `stop` flag, the decision to execute the conditional critical region might be dependent on a value that the thread has read in one of these outer blocks.

3.2.1.4 Condition Synchronization and Nesting

Suppose that we use `retry` to write a synchronization barrier in which a set of n threads waits until they are all ready to continue:

```
// Count threads
atomic {
  threads_present++;
  if (threads_present == n) {
    should_continue = true;
  }
}
// Wait until all threads are present
atomic {
  if (!should_continue) retry;
}
```

The first `atomic` block counts the threads as they arrive at the barrier and sets the flag `should_continue` once all are present. The second `atomic` block waits until the flag is set.

The problem here, as with some of our earlier examples in Section 3.1, is that the code cannot be used if it might occur within a nested `atomic` block. This problem occurs both with closed-nesting and with flattening: the outer level of atomicity prevents the updates to `threads_present` from being visible from one thread to the next. Related problems can occur with nested critical sections or monitors [157].

There are a number of approaches to this problem. One option, of course, is to say that the programmer should not use constructs such as barriers within `atomic` sections: it does not make sense for the complete barrier operation to execute atomically because it naturally involves two separate steps. Type systems or annotations could be used to help avoid this kind of program (e.g., identifying which functions might block internally).

A second option is to provide mechanisms for controlled communication between `atomic` blocks. One possibility are Luchangco and Marathe's *transaction synchronizers* which encapsulate mutable state that is being shared between transactions [203]. Transactions explicitly "synchronize" at a synchronizer. Multiple transactions can synchronize concurrently, whereupon they directly see one another's accesses to the encapsulated data. This can be used, for instance, to produce a TM-aware barrier that allows a set of transactions to proceed only once they have all reached the barrier.

A further form of controlled communication are Smaragdakis *et al.*'s *transactions with isolation and cooperation* (TIC). This system provides block-structured `atomic` actions and a `wait` operation that punctuates them, committing the preceding section and starting a new transaction for the subsequent section. In our example barrier, this would allow the implementation to publish the update to `threads_present`. In addition to `wait`, TIC provides an "expose" construct to identify code sections where this kind of waiting can occur:

```
expose (Expression) [establish Statement]
```

The "establish" statement executes, after waiting, to fix-up any thread-local invariants if the transaction is punctuated by waiting. Static rules ensure that `wait` only occurs under `expose`, and that callers are aware whether or not methods they call may use `expose` and `wait` internally.

3.2.1.5 Condition Synchronization and Race-Freedom

If an implementation of `atomic` blocks provides SLA or TSC for race-free programs, then we must consider exactly which uses of `retry` are considered correct and which are considered to be data races. It is important to have a precise notion of which uses of `retry` are correct because implementations are usually based on instrumenting transactions so that they wake up threads blocked in `retry` (Section 4.6.2)—without a precise definition of which uses are correct, it is not clear exactly where wake-up operations might be needed.

It is not immediately clear how to extend an SLA style of semantics to incorporate synchronization with `retry`: there is no direct counterpart in a lock-based program, and so we cannot simply reduce these constructs to a reference implementation and ask the question whether or not that is race-free.

For an TSC style of semantics, Abadi *et al.*'s notion of *violation freedom* [3] says that a program is incorrect if there is a data race between a normal memory access in one thread and an access in an attempt to run an atomic action in another thread—irrespective of whether the atomic action succeeds normally, or whether it reaches a `retry` statement.

That suggests that the following example is not race-free because of the accesses to x:

```
// Thread 1            // Thread 2
atomic {                x = 42;
  if (!x) {
    retry;
} }
```

Conversely, the following example is race-free because the `atomic` blocks never attempt to read from x under TSC:

```
// Thread 1            // Thread 2        // Thread 3
atomic {               atomic {           x=100
  if (u != v) {          u++;
    if (!x) {            v++;
      retry;           }
} } }
```

Finally, the following example is also race-free because Thread 1 can only read from x inside its atomic block if it executes before Thread 2's atomic block (and, therefore, not concurrently with Thread 2's normal write to x):

```
// Thread 1                    // Thread 2
atomic {                       atomic {
  if (!u) {                      u = 1;
    if (!x) {                  }
      retry;                   x = 1;
} } }
```

3.2.2 EXCEPTIONS AND FAILURE ATOMICITY

Most languages provide some mechanism for non-local control-flow via exceptions, and so it is necessary to consider how this interacts with atomic blocks. There are two main design choices to make: what happens if exceptions are used inside atomic blocks (i.e., raised within a block and handled within the same block), and what happens if an exception is raised within an atomic block, but not handled until outside the block.

The first of these design choices is not contentious: an exception that is raised and handled within a single atomic block should behave as normal. This enables code re-use within atomic blocks. For instance, consider the following example:

```
int x = 1;

try {
   atomic {
      x = 2;
      try {
          throw new Exception();
      } catch (Exception) {
          x = 4;
      }
   }
} catch(Exception) { print(x); };
```

In this case, the exception is raised inside the atomic block and handled by the inner-most catch block. This leaves x with the value 4. The atomic block then finishes normally, without executing the print statement. This design choice reflects the use of exceptions for scoped error handling.

However, suppose that an exception goes unhandled within an `atomic` block. Does this cause the effects of the block to be rolled back? For instance, in the following example, should 1 be printed, or 2:

```
int x = 1;

try {
    atomic {
        x = 2;
        throw new Exception();
        x = 3;
    }
} catch (Exception) { print(x); };
```

Printing 1 corresponds to aborting the transaction when the exception leaves the `atomic` block. Printing 2 corresponds to committing it. There are arguments for each of these choices.

3.2.2.1 Committing

In many ways, it is simpler to commit a transaction while propagating an exception. Committing is consistent with the usual behavior of other kinds of control-transfer: for instance, language proposals typically commit if returning from a function within an `atomic` block. In addition, committing means that, in a sequential program, it is safe to add or remove `atomic` blocks because throwing an exception in an `atomic` block does not change the program's final state.

Committing enables implementations of `atomic` blocks to use lock inference techniques without maintaining information about how to roll back the effects of the block. Furthermore, committing avoids the need to consider *how* to roll back the state manipulated by the `atomic` block while still preserving the exception itself—e.g., if the exception itself was allocated inside the `atomic` block, then it would need to be retained in some form during the roll-back.

3.2.2.2 Aborting

Aborting while propagating an exception enables uses of `atomic` blocks for failure atomicity as well as for isolation. Consider our somewhat contrived example of moving an item between two lists:

```
bool Move(List s, List d, int item) {
    atomic {
        s.Remove(item); // R1
        d.Insert(item); // I1
    }
}
```

Letting an exception at I1 abort the `atomic` block simplifies the code by eliminating the need to provide explicit compensation. With the "committing" semantics, if insert I1 fails, then it is necessary for the programmer to manually undo the effect of R1. More significantly, if a library throws an exception for which a surrounding transaction cannot reasonably be expected to be prepared (out of memory, say, or divide-by-zero caused by a library bug), committing on the way out probably means leaving data structures in an inconsistent state.

This form of failure atomicity has been investigated independently of transactions [160; 286]. Experience often suggests that error-handling code is difficult to write and to debug, and so automating some aspects of state management via roll-back can be valuable, particularly, with low-level failures or asynchronous interruption.

This "aborting" semantics for exceptions agrees with traditional transactional programming—in the first example, the statement x=3 has not executed, and so committing x=2 would be inconsistent with all-or-nothing execution.

3.2.2.3 Discussion

Although the arguments in favor of aborting/committing now seem well understood, there is not a consensus of one approach being strictly better than the other. Hybrid designs are possible, as are techniques to build one form of behavior over the other, so long as some form of user-initiated abort is provided.

One hybrid approach is to distinguish particular kinds of exception that cause roll-back, introducing "committing exceptions" and "aborting exceptions". Alternatively, one might distinguish particular kinds of throw statement, atomic block, or try...catch construct. These alternatives have been explored in transactional programming models for C/C++ [8; 232; 287].

Beyond exceptions, there are further design choices to make in languages that combine atomic blocks with richer forms of non-local control-flow. Kimball and Grossman examined the interactions between atomic blocks and first-class continuations in an extension to Scheme [170]. Additional complexities include continuations captured within an atomic block being resumed after the block has been committed and the use of co-routine-style iteration from within an atomic block.

3.2.3   INTEGRATING NON-TM RESOURCES

As we said at the start of this section, early implementations of atomic blocks lacked orthogonality with other language features: in many cases, it was not possible to perform system calls and IO operations; it was not possible to use non-transactional synchronization constructs; and in some cases, it was not even possible to perform memory allocation reliably. In this section, we try to identify common techniques that have emerged.

Concretely, we look at how to provide integration between atomic blocks and existing binary libraries, storage management, locks, condition variables, volatile data, IO, system calls, and external, non-memory transactions. In each of these cases, our reference point is a semantics that provides atomic blocks with SLA or TSC execution for race-free programs. This kind of reference point lets us untangle two distinct problems:

- First, we must decide how a given language feature ought to behave when used inside atomic blocks. To do this, we consider how it would behave when used in atomic blocks implemented by a global lock or when used inside atomic blocks implemented with TSC.

- Second, we must identify implementation techniques that are consistent with these desired semantics.

With these two problems in mind, there are four basic approaches for handling any particular feature:

Prohibit It. This is appropriate when the semantics of the feature are incompatible with the semantics we want for `atomic` blocks. The barrier synchronization idiom from Section 3.1 is one such example; if a method contains such code, then it may be better to prevent it from being called inside `atomic` blocks and thereby detect possible problems at compile-time.

Execute It. Some abstractions can simply be instrumented with `ReadTx`/`WriteTx` operations and executed as part of a transaction—e.g., this applies to many standard libraries.

Irrevocably Execute It. A general-purpose fall-back technique for IO and system calls is to use *irrevocable* execution (also known as *inevitability*). When a transaction becomes irrevocable, the TM implementation guarantees that it will not be rolled back. This allows it to execute non-speculatively. In many implementations, the thread running an irrevocable transaction can invoke normal non-transactional operations without needing to consider how to roll these back or compensate for them. Forms of irrevocable execution have been explored both in software (e.g., [303; 333], Section 4.6.3) and in hardware (e.g., [37; 197]).

Integrate It. The final general technique is to integrate an abstraction with TM. This may mean writing a special version of the abstraction for use inside `atomic` blocks and providing transaction-aware logging, compensation, and conflict detection for threads that are using the abstraction transactionally [50; 133; 190; 215; 228; 243; 322; 346; 346]. For instance, a TM-integrated error-logging facility might batch up log entries in an in-memory buffer before flushing them to disk when the transaction commits. The buffer can simply be discarded if the transaction aborts. As with irrevocability, such techniques are applicable to hardware-based implementations, as well as software.

Examples. Having introduced these four main techniques for integrating non-transactional resources, we now consider a series of examples showing how the techniques can be applied in practice (Sections 3.2.4–3.2.7).

3.2.4 BINARY LIBRARIES

Libraries are an integral part of any software, and so `atomic` blocks need to call precompiled or newly compiled libraries. For full orthogonality, programmers might expect to use features like dynamic linking and loading, or to combine libraries from different compilers and languages into a single application.

In the longer term, some of this kind of interoperability may come through standardization on common TM implementations, common calling conventions for transactional functions, and so on. In the shorter term, dynamic binary rewriting or switching to irrevocable execution on non-transactional libraries provides a fall-back.

3.2.5   STORAGE ALLOCATION AND GC

Storage allocation and GC both pose challenges for use within `atomic` blocks. These operations provide examples of why many abstractions do not simply "just work" if they are instrumented with `ReadTx`/`WriteTx` calls for use in TM.

Explicit Allocation and De-Allocation.   Starting with allocation, part of the problem is that storage allocation is used both by transactional and non-transactional code. Even if the TM supports strong isolation, then there is still the question of how to manage allocations that involve system calls to expand the heap or that trigger access violations to demand-initialize new pages.

   To handle these problems, many implementations of `atomic` blocks use integration between storage allocation and the TM runtime system. Commonly, allocations execute non-transactionally; if the transaction commits, then this tentative allocation is retained. De-allocation requires more care. It is not always sufficient to defer de-allocation requests until a transaction commits. For instance, consider this contrived example:

```
void TemporaryData() {
  atomic {
    for (int i = 0; i < 1000000; i ++) {
      void *f = malloc(1024);
      free(f);
    }
  }
}
```

Executed without the `atomic` block, this would run in constant space, allocating and deallocating a series of blocks of memory. If it is to do the same with the `atomic` block, then the implementation must track the status of memory allocations and identify *balanced* de-allocations, in which data is allocated and then de-allocated within a single transaction [162].

   In addition to tracking balanced usage, a second problem with de-allocation is deciding when the storage becomes eligible for re-use. The recycling of memory from transactional usage, through the allocator, to non-transactional usage, can give rise to privatization-style problems: one thread might be accessing the data speculatively within a zombie transaction, even though another thread has re-allocated the data for some other non-transactional use. Consequently, in systems where zombie transactions can occur, non-transactional use of a given memory block must be delayed until any speculative transactional use has finished—for instance, by delaying re-use until any transactions preceding the de-allocation have been validated (Section 4.6.1). Similar re-use problems are faced in many systems using lock-free synchronization [105; 144; 222], and they led to epoch-based lists of tentative de-allocations that are awaiting re-use. This kind of problem is also one of the central issues addressed by RCU [217].

Garbage Collection.   It is rarely practical to combine garbage collection with TM by simply instrumenting the collector and running it in the context of whatever transaction happens to trigger GC. The volume of data accessed would be vast, and the synchronization used during GC may be incompatible with the TM. In practice, the GC is really part of the language runtime system, and so it should be integrated with the TM rather than being executed transactionally.

Some early implementations of atomic blocks simply aborted all transactions when running GC [134]. This seems reasonable if atomic blocks are short or GC comparatively sporadic. However, it is less reasonable if atomic blocks are long-running and might need to span a full GC cycle in order to complete successfully. In any case, integration between the GC and TM must be explored in settings that use incremental GC (in which small amounts of GC work are mixed with application work) or concurrent GC (in which the GC runs concurrently with application code).

A basic approach is to treat the TM system's read/write-sets as part of the GC's roots, either ensuring that the objects they refer to are retained by the collector or updating them if the objects are moved during collection [136]. With this scheme, if GC occurs during a transaction that subsequently aborts then there is no need to revert the garbage collection or to put moved objects back in their old locations: in effect, the TM operates above the level of the GC. It is less clear how to exploit this approach in an HTM system in which a software GC would typically operate above the level of the TM.

As Shinnar *et al.* [286], Harris *et al.* [138] and Detlefs and Zhang [81] discuss, integration between TM and a GC can introduce some complexity if the TM logs refer to *interior* addresses of individual fields, and the runtime system does not provide a ready mechanism to map these addresses back to object references.

Additional care is needed to support variants of the TemporaryData example: if a transaction allocates temporary data structures, then references to these can build up in the transaction's logs, keeping the temporary objects alive even when the application has finished with them.

The approach taken with Bartok-STM [138] is to arrange for the collector to consider two views of the heap: one in which all current transactions are rolled back and another in which all current transactions are committed. Objects are retained if they are reachable in either one of these views. For simplicity, all active transactions are validated before doing this, and any invalid transactions are rolled back immediately. This means that the remaining transactions have non-overlapping sets of updates, so there are at most two values for any given field.

TM can also be useful in the implementation of GC. As McGachey *et al.* show, this is particularly apparent in the case of systems where GC work proceeds concurrently with the execution of an application [216]: transactions can be used to isolate pieces of GC activity, such as relocating an object in memory.

Class Initialization, Object Finalizers, and Destructors.   The final storage-related question is how to run initialization and clean-up code such as static initializers and finalizers in Java or constructors and destructors in C++.

In Java, the language specification is very precise that class initialization occurs exactly once at the first use of each class. If this use occurs within an atomic block, then the language designer must consider whether the initializers executes transactionally (and may, therefore, be rolled back if the transaction aborts) and whether or not it sees any tentative updates of the transaction that triggers loading. The implementation of class initialization is often complicated and involves native parts of the JVM implementation. Irrevocable execution may be the only pragmatic option.

In terms of clean-up during object de-allocation, then the first question is if operations such as finalizers and destructors should run at all? Under SLA or TSC, transactions that abort due to conflicts are not visible to the programmer, so arguably the objects allocated during these transactions should appear to never have been allocated. Conversely, it seems clear that finalizers and destructors should run normally for objects that have been allocated by transactions that ultimately commit.

It is not clear how to handle objects in examples like the `TemporaryData` one where data is allocated and destroyed within a transaction. For instance, does the finalizer run within the context of the transaction itself or within some other thread granted special access to the transaction's tentative work?

It is possible that examples like these do not occur in practice, and so the use of finalizers within transactions could be prohibited—typical applications of finalizers involve coordinating de-allocation of external resources, and these cases are better handled at a higher level by proper integration with TM.

3.2.6 EXISTING SYNCHRONIZATION PRIMITIVES

Should existing synchronization primitives work within `atomic` blocks, and if so, how?

As Volos *et al.* illustrate, it is usually insufficient to instrument an actual *implementation* of a synchronization primitive to use `ReadTx/WriteTx` operations and to expect things to work out correctly [321]. One option, therefore, is to prohibit the use of these primitives [136; 242]. However, this does prevent re-use of existing libraries that might use locks internally—and locking is pervasive in many languages' standard libraries.

If a system is to support synchronization primitives inside `atomic` blocks, then we can use models such as SLA and TSC to define the semantics of the constructs in a methodical way, and then investigate whether or not practical implementation techniques are available. In this section, we briefly consider combinations of `atomic` blocks with locks, condition variables and `volatile` fields in a Java-like language. We focus on Java's form of `volatile` fields in which accesses to the fields execute with sequential consistency.

3.2.6.1 Locks
Semantically, accommodating locks within `atomic` blocks is relatively straightforward. Consider the following example:

```
// Thread 1                        // Thread 2
atomic {                           synchronized(obj) {
  synchronized (obj) {               x++;
    x++;                           }
  }
}
```

Under both SLA and TSC, the two accesses to x cannot occur concurrently because the threads performing them would both require the lock on obj; one would expect the result x==2. In other examples, there are subtle differences between SLA and TSC. Consider this example using nested critical sections:

```
// Thread 1                        // Thread 2
atomic {                           synchronized(obj) {
  Object n = new ...                 temp = x;
  synchronized(obj) {              }
    x = n;                         // Use temp
  }
  // L1
}
```

Under TSC, Thread 2's synchronized block either executes entirely before Thread 1's atomic block or entirely after it. This is because TSC semantics prohibit any other thread's work from appearing to be interleaved with an atomic block's execution. An implementation could achieve this by deferring Thread 1's unlocking of obj until the enclosing atomic block commits.

However, under SLA, Thread 2's synchronized block is permitted to run concurrently with Thread 1's execution of L1: Thread 2 requires the lock on obj, but not the lock conceptually held by Thread 1's atomic block. Furthermore, Thread 2 would be expected to see the work done in Thread 1's synchronized block and in the preceding part of Thread 1's atomic block. To support these semantics, it may be necessary for Thread 1's transaction to become irrevocable just before the point where the lock on obj is released. This example illustrates the subtleties that can occur when mixing transactions and other constructs and the need for precise specifications of the behavior that is intended.

Finally, consider the example below:

```
// Thread 1                        // Thread 2
atomic {                           synchronized(obj) {
  x = 42;                            x = 42;
}                                  }
```

This is simply a data-race under TSC and under SLA: although both accesses are inside some kind of synchronization construct, there is nothing to prevent them being attempted concurrently (much like two threads accessing the same data while holding different locks from one another). Some systems do aim to support this kind of mixed-mode synchronization by preventing transactions running concurrently with lock-based critical sections or by using programmer-annotations to control concurrency [111].

3.2.6.2 Condition Variables
Condition variables pose a particular problem within atomic blocks. Signaling a condition variable (notify/notifyAll in Java) is straightforward: it can be deferred until the atomic block commits.

The difficulty comes from a `wait` operation. For instance, consider the typical usage when waiting to update a shared data structure:

```
atomic {
  ...
  synchronized (buffer) {
    while (full==true) buffer.wait();
    full = true;
    item = i;
  }
  ...
}
```

Ordinarily, the section of code before the `wait` should appear to run *before* the condition variable has been signaled, but the section of code after the `wait` should appear to run *after* the signal. The problem here is the same as the barrier example in Section 3.1: when both of these sections of code involve side-effects, exposing one set of effects and not the other is incompatible with the entire block running atomically. The problem with condition variables is therefore more one of defining semantically how they should interact with `atomic` blocks, rather than just one of implementation complexity.

Various definitions are possible. Under SLA, behavior must be equivalent to using a single lock in place of the `atomic` block: the nested `wait` would release the lock on the buffer but not the enclosing lock used (conceptually) by the `atomic` block. Work preceding the `wait` should be visible to another thread that acquires the lock on the buffer. Under TSC, no other threads' work should appear to be interleaved with the `atomic` block.

Under either semantics, it may be appropriate to prohibit the use of `wait` within `atomic` blocks, or to warn the programmer of the risk that they will write a deadlock in this way. If an abstraction that uses condition variables internally is to be used inside `atomic` blocks, it would be necessary to integrate the abstraction with TM at a higher level—for instance, providing a variant of a buffer with TM-aware synchronization, rather than condition-variable synchronization.

From a practical viewpoint, Carlstrom *et al.* examined the behavior of Java applications built using TM [49]. They observed that waiting within nested `synchronized` blocks is generally discouraged in any case, and so one might expect that `wait` would typically not be used within `synchronized` blocks within `atomic` blocks. They also saw that treating `wait` as a `retry` allowed many idioms to work, in effect only running an `atomic` block when it can complete without calling `wait`.

Other approaches are possible. AtomCaml provides a form of condition variable in which a thread finishes part of an atomic action, committing its work thus-far, then waits, and then starts a new atomic action. Ringenburg and Grossman note how the thread must start listening for notifications when it commits the first atomic action, to avoid lost-wake-up problems [266]. Dudnik and Swift discuss implementing this form of waiting in a C-like language [95], splitting a normal `wait` operation into a "prepare wait" step to execute inside an `atomic` block, and a "complete wait" step for execution outside the block.

3.2.6.3 Volatile Data

In Java and C#, `volatile` data provides a low-level synchronization mechanism between threads. Concurrent accesses to `volatile` fields *do not* form data races in these languages, and so a program using `volatile` data would be expected to be race-free under SLA and under TSC. In addition, neither Java nor C# allows `volatile` memory accesses to be reordered with one another.

In rather ugly code, one might write:

```
volatile int x;

// Thread 1            // Thread 2
atomic {               r1 = x;
  x = 10;              r2 = x;
  x = 20;
  x = 30;
}
```

Under SLA, the reads into `r1` and `r2` may see a series of values for x, so long as they are consistent with the order of the writes from Thread 1—e.g., `r1==10, r2==20`, but not `r1==20, r2==0`. Under TSC, the reads should not see intermediate values from the `atomic` block, so only 0 and/or 30 should be seen.

Providing either of these semantics requires care if multiple `volatile` fields are involved, or if an update to a `volatile` field is used to publish a transactionally-allocated data structure. It may be much simpler to use irrevocability if cases like this are rare.

3.2.7   SYSTEM CALLS, IO, AND EXTERNAL TRANSACTIONS

Perhaps, the most serious challenge in using transactions is communication with entities that are outside the control of the TM system. Modern operating systems such as Unix and Windows provide a very large number of mechanisms for communication, file manipulation, database accesses, interprocess communication, network communication, etc. The implementations of these operations typically involve making system calls to the OS, and the use of these operations can be buried deep within libraries. These operations involve changes in entities which are outside the control of the TM system: if atomicity is to be extended to include these operations then mechanisms are needed to prevent conflicts between concurrent operations and to roll back state if a transaction needs to be aborted.

While there is no general mechanism to manage these operations, ideas such as SLA or TSC can be used to define the intended behavior of different operations in a consistent way. As we discuss below, many implementation techniques exist on a case-by-case basis, and switching to irrevocable execution often provides a fall-back [37; 303; 308; 333].

A system can buffer operations such as file writes until a transaction commits and then write the modified blocks to disk. Similarly, the system could also buffer input, to be replayed if the transaction aborted and reexecuted. These seem like simple solutions until we combine them in a transaction that both reads and writes. Suppose the transaction writes a prompt to a user and then waits for input. Because the output is buffered, the user will never see the prompt and will not produce the input. The transaction will hang.

In some cases, compensating actions can be used to undo the effects of a transaction. For example, a file write can be reverted by buffering the overwritten data and restoring it if a transaction aborts. Compensation is a very general mechanism, but it puts a high burden on a programmer to understand the semantics of a complex system operation. This becomes particularly difficult in the presence of concurrency when other threads and processes may be manipulating the same system resources (e.g., in the case of the file write, then the tentatively-written data must remain private until the writing transaction commits).

Many systems support transaction processing monitors (TPMs) [31], which serve as a coordinator for a collection of systems, each of which supports transactions and wants to ensure that the operation of the collection appears transactional to all parties outside the group. TPMs generally use a two-phase commit protocol, in which the constituent transactions first all agree that they are ready to commit their transactions (and all abort if any wants to abort) and then commit *en masse*.

A number of papers discuss transaction handlers that invoke arbitrary pieces of code when a transaction commits or aborts [50; 133; 215; 346], or which provide for two-phase commit between different transactional constructs [133; 190; 215]. These handlers can interface transactions to TPMs or other transactional systems and implement compensating actions to revert external side effects.

Baugh and Zilles studied the extent to which different techniques apply to the system calls made by large applications [29] and Volos *et al.* developed a TM-aware API for handling many resources [322]. They introduce "sentinel" objects which represent associations between threads and logical state within the kernel (e.g., access to a given file). These sentinels are used to signal conflicts to the transactions involved and avoid the need to hold long-term locks on data structures within the kernel.

Another approach is to allow IO operations in transactions only if the IO supports transactional semantics, thereby enabling the TM system to rely on another abstraction to revert changes. Databases and some file systems are transactional, and there is renewed research interest in broadening support for transactional styles of interaction [243; 244] (in part motivated by enabling integration into transactional programming models). However, the granularity of these systems' transactions may not match the requirements of an atomic block. For example, Windows Vista supports transactional file IO, but these transactions start when a file is first opened. Therefore, if an atomic block only performs one write, it is not possible to use a file system transaction to revert this operation, without discarding all other changes to the file.

3.3   PROGRAMMING WITH TM

In this section, we briefly survey the practical experience that has been gained in programming with prototype implementations of TM and `atomic` blocks—for instance, integration with debuggers and profilers (Section 3.3.1), the development of benchmark workloads (Section 3.3.2), and user studies of programmer productivity (Section 3.3.3).

3.3.1   DEBUGGING AND PROFILING

Software development tools and development environments must evolve to support `atomic` blocks and TM. If an SLA semantics is provided, then this can allow the debugging of transactional applications to be reduced to the problem of debugging lock-based programs—either by directly executing the program using locks or by using the hypothetical behavior of the program using locks as a guide for the behavior of the debugger when using transactions.

More broadly, however, the concepts of optimistic or aborted execution do not exist in today's tools. What does it mean to single step through an atomic transaction? A breakpoint inside the transaction should expose the state seen by the transaction. However, how does a debugger present a consistent view of a program's state, since part of the state not relevant to the transaction but visible to the debugger, may have been modified by other threads? Furthermore, with lazy version management, transactionally modified state resides in two places: the original location and a buffered location. The debugger must be aware of this separation and able to find appropriate values. In addition, how does a programmer debug a transaction that aborts because of conflicts? Lev and Moir discuss the challenges of debugging transactional memory [192].

In more recent work, Herlihy and Lev describe the design and implementation of a debugging library for programs using TM [143]. Their system "tm_db" is designed to provide a common interface that, from one side, can be used by a debugger and, from the other side, can be supported by a range of TMs. They propose that, from the point of view of the programmer, only the *logical value* of a memory location should usually be presented—e.g., if a TM implementation uses eager version management, then these tentative operations should remain hidden until a transaction commits. When single-stepping, the debugger shows the logical values for most of the heap and shows non-committed state only from the current thread.

Zyulkyarov *et al.* proposed similar ideas, letting a user distinguish between this kind of "atomic-level" view of the program, in which the user should not have to be aware of the details of how the blocks are implemented (e.g., TM, or lock inference), and a "transaction-level" view, in which a user investigating lower level problems can investigate transaction logs, conflicts, and so on [351]. Zyulkyarov *et al.* also introduced mechanisms to allow transactions to be introduced or removed from within the debugger, for instance, allowing a series of operations to be dynamically grouped into a transaction, to allow the user to investigate whether or not a bug is occurring due to a lack of atomicity between the operations.

In communicating with the user, Herlihy and Lev propose distinguishing the *logical trans-action*, which represents a single attempt to execute an `atomic` block, from a series of *physical transactions* which are used to implement it. Thus, a transaction identified as 3.4.1 might indicate thread number 3, executing logical transaction 4 for the 1st time. This helps indicate progress to the user, and it provides an ID scheme that can be understood in terms of the source program.

Aside from debugging, profilers and other tools are important both for the uptake of TM-based programming models and for developing a deeper understanding of the characteristics of transactional workloads.

Elmas described how a data-race detector for lock-based programs can be extended to accommodate synchronization using transactions [98]. It is based on a *happens-before* ordering between synchronization operations and can be configured to support an SLA-style model in which this is a total ordering between the start and commit of all transactions or a DLA-style model in which only transactions accessing overlapping data are ordered. As in our examples in Section 2.2, this leads to different notions of which programs are considered race-free.

At a coarse granularity, several researchers have studied whole-program characteristics, such as commit/abort ratios, the sizes of read/write-sets, or histograms of how frequently different transactions need to be re-executed [17; 65; 241]. This provides some guidance for tuning TM implementations – for instance, what volumes of data need to be handled in the common case – but only limited insight into exactly where a problem might occur within a large application that uses transactions in multiple ways.

Chafi *et al.* developed finer-grained profiling hardware for the TCC HTM implementation [54] (Section 5.3.3). Their system, "TAPE", tracks which transactions overflow hardware buffers and which transactions and data are involved in conflicts. Zyulkyarov *et al.* described a "conflict point discovery" profiling technique, based on identifying the frequency with which operations at different PCs trigger conflicts [351].

3.3.2   TM WORKLOADS

Many early papers on transactional memory used simple data-structure microbenchmarks such as linked-lists, red-black trees, and skip-lists. This reflects one possible use for transactions: enabling higher-performance implementations of this kind of nonblocking data structure. Such workloads may involve transactions making a handful of memory accesses, and it may be reasonable to assume that they are written by expert programmers.

More recently, many TM researchers have investigated writing larger parts of applications using transactions. These workloads may involve transactions containing hundreds (or even thousands) of memory accesses. This second kind of TM workload often reflects a use of transactions in the hope that they are easier to use than existing synchronization constructs, rather than because the implementation is believed to be faster than existing alternatives.

3.3.2.1   Synthetic Workloads

STMBench7 [127] is a TM-based benchmark developed from the OO7 benchmark for object-oriented databases. STMBench7 provides coarse-grained and medium-grained lock-based versions, along with transactional versions for various languages. The benchmark performs complex and dynamically-selected operations on a complex pointer-based data structure.

WormBench[349] is a synthetic application implemented in C#, providing a configurable level of contention and configurable read/write-set sizes. It can be used to mimic characteristics of other applications to study or debug some performance issues in a simpler setting.

3.3.2.2 Applications and Computational Kernels

The Stanford Transactional Applications for Multi-Processing benchmark suite (STAMP) provides eight applications that have been structured to use transactions [47]. The STAMP workloads exhibit a range of characteristics in terms of the time that they spend inside transactions (from almost none, up to almost 100%) and in terms of the size and conflict characteristics of the transactions (from very short transactions to those spanning hundreds of thousands of instructions). Kestor *et al.* developed transactional versions of a set of "Recognition, Mining and Synthesis" applications developed from BioBench and MineBench [168].

In addition to general benchmarks suites, a number of TM workloads have illustrated particular uses for TM or particular synergies between TM and different forms of algorithm:

Scott *et al.* examined the use of TM in implementing a Delaunay triangulation algorithm [283]. Aside from providing an example TM workload, this program demonstrates how data can change between being thread-private to being accessed transactionally. Programming models where transactional and non-transactional data are statically separate would not work well here, and any overhead on the non-transactional accesses would harm overall performance.

Kang and Bader designed a TM-based algorithm to compute a minimum spanning forest of sparse graphs [167]. They argue that graph-based algorithms are well suited to TM because such algorithms can involve visiting a dynamically selected series of nodes (making it difficult to lock nodes in advance), and many algorithms can be expressed as atomic operations on a single node and its neighbors (thereby making the size of transactions small and a good fit even for bounded-size TMs). They show how an algorithm for finding minimum spanning trees can operate as a step-by-step graph exploration algorithm of this kind.

Nikas *et al.* investigated the use of TM in an implementation of Dijkstra's algorithm for computing single-source shortest paths [16; 233], showing how optimistic synchronization enables helper threads to opportunistically process non-minimum nodes from a priority queue.

3.3.2.3 Games

Spear *et al.* [308] used part of a 3D game to evaluate the use of irrevocable transactions. Their system involves animating a set of multi-segment objects and performing collision detection between them. The game can use irrevocable transactions to perform periodic rendering of the current state of all of the objects, thereby allowing the rendering to proceed regularly and to be decoupled from the underlying physical simulation.

Zyulkyarov *et al.* [350] and Gajinov *et al.* [107] each investigated techniques for expressing parallel implementations of the Quake game server using TM, investigating where `atomic` blocks work well, and where they do not. The main positive results came from the core game data structure representing the map. Encapsulating a player's move in a transaction simplified the logic when compared with a lock-based implementation, removing the need for a preparation phase that was used to select which parts of the map to lock. TM worked less well in cases where there were external interactions and in cases where the structure of the code did not fit with block-structured transactions. Both of these observations point to potential difficulties when incrementally introducing transactions into an existing program.

Baldassin and Burckhardt explored the use of TM to parallelize a simple "Space Wars 3D" game [27], finding the basic programming model convenient, but performance hampered both by high levels of contention and high sequential overhead. They improved performance by allowing transactions to run concurrently, working on private copies of objects and introducing explicit *merge functions* between the states of objects that have experienced a conflict. For instance, simply picking one update over the other, or combining them if they commute.

3.3.2.4 Programming Paradigms

Singh developed a library of *join patterns* using `atomic` blocks in STM-Haskell [294]. These patterns provide a form of synchronous interaction between a thread and a set of channels; for instance, waiting for an input value to be available from both of two channels. The `orElse` construct provides a natural way to combine alternatives, and the `retry` construct provides an opportunity to wait for a value which matches a user-supplied predicate; if a value putatively received is not valid, then `retry` abandons the current transaction and waits for an alternative value to be available.

Lam and Sulzmann developed a library for constraint logic programming using the GHC Haskell implementation of `atomic` blocks [185]. The system maintains a "constraint store" and uses a set of domain-specific rules to simplify it (e.g., to compute greatest-common-divisors, a rule might replace `Gcd(m)` and `Gcd(n)` with `Gcd(m-n)` and `Gcd(n)`). Lam and Sulzmann model each rewrite rule by a thread and use STM to manage the constraint store to prevent conflicting updates. To avoid conflicts on a linked-list used to store the constraints, threads take "small steps" along the list in separate transactions, and then they manually re-validate these by additional validation when making an update. Perfumo *et al.* study the performance of this system on a range of problems [241].

Donnelly and Fluet [91] and Effinger-Dean *et al.* [96] designed transactional-event systems that build on message-passing programming paradigms. If a set of transactions communicates with one another, then their fates get bound together, in the sense that they must either all commit or all abort. Ziarek investigated the use of *memoization* in this kind of setting, tracking the communication performed by a transaction, and allowing the results of an earlier communication to be re-used if the operation that triggered it is re-executed [343; 344].

3.3.2.5 TM in the OS kernel

Rossbach *et al.* explored workloads within the Linux operating system kernel for suitability for implementation with TM [253; 270; 271]. This provides a large-scale test of how TM performs in a real system.

They developed a special-purpose co-operative transactional spinlock (`cxspinlock`) for use in this setting. Ordinarily, a `cxspinlock`'s holder executes transactionally and detects conflicts with other threads based on the data that they are accessing (somewhat like a software implementation of Rajwar and Goodman's SLE system [248], Chapter 5). A `cxspinlock`'s implementation selects dynamically and automatically between locks and transactions, starting out with transactions and falling back to locking if the kernel attempts an IO operation. Transactional use of the `cxspinlock` is prevented while any thread holds it as an actual lock; this prevents problems with mixed transactional and non-transactional accesses. This approach avoids the need for system-wide irrevocable transactions; at worst, the system scales according to the placement of the locks.

3.3.2.6 Discussion

Benchmarks such as the STAMP suite provide a common starting point for much work on evaluating the performance of transactional memory, and, inevitably, experience will grow about how transactions are best used and where they perform well and where they do not. However, even as they stand, the current benchmarks provide some indications of recurring patterns for using TM and how programming models can affect workloads.

A common pattern in many of these benchmarks is to use optimistic concurrency control to execute items from a shared work list. Transactions are used to serialize independent items, allowing concurrency when the contention rate is low, but providing as-if-sequential behavior if conflicts exist. The data structures accessed by the work items are often complicated (so designing fine-grained locking would be difficult), or there are many data structures that are accessed together (so locking could not be encapsulated within them), or the potential access set is vastly lager than the actual set touched (e.g., in Labyrinth from the STAMP suite). Any of these properties makes it difficult to design an effective scalable locking implementation. In addition, if transactions are formed from items from a work queue, then serializing these items directly provides an effective fall-back technique when there are few cores or when contention is high.

3.3.3 USER STUDIES

Finally, we turn to user studies that try to address the question "*Is transactional programming actually easier?*" There are a few initial results:

Rossbach *et al.*'s study examined 147 undergraduate students who had been set a series of programming challenges using fine-grained locks, coarse-grained monitors, and TM [269]. In each case, the problem was to model access to different lanes in a "shooting gallery" game. Broadly speaking, the syntax of the initial TM implementation was found to be cumbersome (more resembling the explicit interface of Chapter 2 than language-integrated `atomic` blocks). Subjectively, transactions were found more difficult to use than coarse-grained locking, but they were easier than developing fine-grained locking. Synchronization errors were reduced when using transactions—moderately so when compared with coarse grained locking, but vastly so with fine-grained.

Pankratius *et al.* studied a group of 12 students working on a parallel desktop search engine in C [238]. The projects were built from scratch, performing indexing and search functions. The students were initially introduced to parallel programming and the TM prototype. Three pairs used a combination of pthreads and locks, and three pairs used pthreads and `atomic` blocks. The execution time of the best TM-based implementation of the indexer was around 30% of the time of the best lock-based implementation and faster than locks on 9 out of 18 of the queries.

3.4   ALTERNATIVE MODELS

The combination of `atomic` blocks and TM occurs so frequently that the two concepts are often combined or confused. In this section, we describe some of the alternatives which illustrate different language constructs that can be built over TM.

One alternative, going back to some of the earliest work on STM [146] is to use a library-based interface to transactions, rather than language extensions. An advantage of a library-based approach is that it can operate without requiring changes to a compiler. Gottschlich *et al.* describe the design of a library-based interface to TM from C++ [108; 110]. Dalessandro *et al.* have argued that a carefully-designed interface can be a significant improvement over the kind of simple explicit interface we used in Chapter 2—for instance, C++ smart pointers can be used to remove some of the verbosity of explicit method calls for each memory access [75]. Nevertheless, where language extensions are possible, Dalessandro *et al.* observed that they led to clearer programs, and they can provide greater decoupling between the code written by a programmer and the details of a particular TM implementation.

3.4.1   TRANSACTIONS EVERYWHERE

Several research projects have investigated programming models where atomicity is the default. These "transactions everywhere" [184] models have an intuitive appeal: start with a correct, simple program built from large atomic actions, and then refine these to smaller atomic steps if this is necessary to achieve good performance.

The TCC programming model is an early example of this approach [131], combining innovative language features with a new underlying memory subsystem (we return to the hardware components in Section 5.3.3). When programming with TCC, the programmer divides the program coarsely into atomic actions that run concurrently on different processors. The TCC implementation ensures atomicity and isolation between these blocks. The programmer can also constrain the commit ordering of the blocks, if this is important.

For example, consider the following sequential program to compute a histogram in `buckets` of the number of times that values are seen in the `data` array:

```
for (i = 0; i < 1000; i++) {
   buckets[data[i]]++;
}
```

By default, this will run as a single atomic action. The loop can be parallelized by replacing it with a `t_for` construct, making each iteration into its own atomic action and guaranteeing that the iterations will appear to occur in the same order as the sequential code. In this case, since the iterations' work is commutative, it would be safe to use a `t_for_unordered` loop which allows the iterations to commit in any order. Further variants allow iterations to be chunked to reduce overheads. While-loops, explicit `fork` operations, and lower-level synchronization constructs are also available.

With the Automatic Mutual Exclusion (AME) model, the program's `main` method is treated as a single atomic action, and explicit `yield` statements are included to indicate points at which atomicity can be punctuated [1; 2; 6]. An asynchronous method call operation is provided to allow one atomic action to spawn another; the new operation will occur after the atomic action it was spawned within. A simple implementation of AME could run these asynchronous actions sequentially, whereas a TM-based implementation could speculatively run them in parallel in separate transactions. A `BlockUntil` operation is provided to allow an atomic action to delay its execution until a given predicate holds. For example, an atomic action may wait for a queue to become non-empty before removing an item from it:

```
BlockUntil(queue.Length() > 0);
data = queue.PopFront();
```

Combining these features, one could write a loop to process different items from an array in parallel:

```
for (int i = 0; i < a.Length ; i ++) {
   async ProcessItem(i);
}
yield;
BlockUntil(done == a.Length);

...

void ProcessItem(int i) {
   ...
   done++;
}
```

In this example, the initial `for` loop atomically spawns separate asynchronous actions to process each element of the array. The `yield` operation completes the atomic action containing the loop, and it allows the asynchronous processing to occur. The `ProcessItem(i)` method handles item `i` and increments the count of items that have been finished. Finally, the code after `yield` starts a new atomic action, and it delays it until the processing is finished. With AME, one could imagine starting with this as a simple, "clearly correct" program, and then relaxing atomicity by splitting

`ProcessItem` into a larger number of small asynchronous work items (to use more cores), or adding additional `yield` points.

A downside of this kind of approach is that programmers must be careful of whether or not `yield` is used in functions that they call. For instance, if the author of `ProcessItem` wants to make sure that it occurs as a single atomic action, then he or she must take care not to call into any functions that might `yield`. To clarify this, AME requires that the programmer identify which functions might yield, both at the function's definition and at all of its call sites.

Finally, the Concurrent Objects with Quantized Atomicity system (COQA) [198] provides an object-oriented style of atomicity which draws, in particular, on Actor programming models. In COQA, methods operate as atomic tasks, and an implementation can be built using TM-like techniques to allow a task to dynamically "acquire" the objects that the method uses. The programmer, therefore, does not need to be concerned about interleaving between individual operations in methods on the same object. COQA provides three kinds of method call to express possible concurrency or relaxation of atomicity. In a normal synchronous call (`o.m(v)`), the callee executes in the same atomic task as the caller. An asynchronous call (`o->m(v)`) allows a new task to be created; this task operates atomically, but is independent of the caller. Finally, a sub-task operation (`o=>m(v)`) is provided to express a form of synchronous call akin to open nesting: the callee executes synchronously, but any resources it acquires are released at the end of the call, and become available to other tasks.

3.4.2   LOCK-BASED MODELS OVER TM

An alternative to designing new language constructs which use TM is to develop techniques that use it to implement existing features—in effect, using TM as an implementation technique, rather than a programming interface [42].

The most popular technique is to implement ordinary critical sections using TM instead of using locks. This approach has two attractions: it can improve the scaling of existing programs without requiring code changes, and it avoids the need to design new language constructs for exposing TM.

With *speculative lock elision* (SLE) [248], critical sections execute speculatively with TM-like techniques being used to dynamically detect conflicts between them. If there is a conflict between speculative critical sections then one or other of the critical sections can be re-executed, or the implementation can fall back to non-speculative execution and actually acquire the lock in question. Conflicts between speculative and non-speculative critical sections can be detected by having speculative sections monitor that the locks they acquire are currently available.

Rajwar and Goodman's original hardware design for SLE [247; 248] transparently identified instruction sequences that were idiomatic of lock acquire/release operations. We return in detail to the hardware mechanisms used by SLE in Chapter 5. Software implementations have been integrated in language runtime systems, based on using TM-style implementation techniques to modify the reads and writes being performed in speculative sections [272; 330; 345].

Care is needed to ensure that an implementation of locks over TM is correct. Programming idioms such as the privatization and publication examples are race-free when implemented with locks, and so these must be supported when locks are implemented with TM. In addition, as Ziarek *et al.* discuss [345], to allow completely transparent interchange between locks and transactions in Java, the set of schedules induced by transactional execution must be a subset of the set of schedules permitted under locks as defined by the Java Memory Model [207]. By framing the development in this way, it is possible to handle programs even if they are not race-free.

An additional source of problems when using TM to implement lock-based critical sections is how to handle nesting. Consider this example, which illustrates how Java's `synchronized` blocks cannot be transparently replaced by `atomic` sections:

```
// Thread 1                   // Thread 2
synchronized (l1) {           while (true) {
  synchronized(l2) {            synchronized(l2) {
    x=10;                         if (x == 10) {
  }                                 break;
  while (true) {                  }
  }                             }
}                             }
```

In this example, Thread 1 writes 10 to x inside a critical section. Meanwhile, Thread 2 continually reads from x, and breaks out of its loop when it sees 10. The location x is consistently protected by lock l2. When implemented with locks, it is not possible for both of these threads to get stuck looping forever, and so this should not be possible with a TM-based implementation of the `synchronized` blocks.

However, Thread 1's operations are wrapped inside an enclosing `synchronized` block protected by lock l1, and so a naïve implementation of `synchronized` blocks using TM could keep *all* of Thread 1's work buffered speculatively. Implementing this with TM is not as simple as separating out Thread 1's accesses under l1 and l2: using locks it would be entirely correct for the thread to perform operations between its two lock acquires, and for Thread 2 to expect to see these.

Building on work by Welc *et al.* [330], Ziarek *et al.* [345] investigated these interactions in detail, developing an execution environment that combines lock-based and TM-based implementations of `atomic` blocks and `synchronized` regions. They term this approach "pure-software lock elision" (P-SLE) since it operates on conventional hardware. The runtime system dynamically detects whether the semantics provided by the TM system are compatible with the use of the language constructs, in which case it switches to a lock-based fall-back implementation.

Conceptually, the challenges in using TM for critical sections like these are simpler than the challenges of designing semantics for `atomic` blocks. First, in this case, the existing semantics of the synchronized blocks provide a definition for what the implementation should do, and avoid the ambiguity over whether a particular example should be considered a badly-synchronized program or whether a particular implementation technique should be considered incorrect.

Second, in any problematic cases, the implementation can fall back to non-speculative execution and directly acquire the locks—nesting, IO, and the use of other synchronization techniques can all be handled in this way. This permits a range of implementation techniques, including simple ones where locks are used frequently and more complex ones that can execute more programs speculatively.

Usui *et al.* investigate a hybrid adaptive-locking approach, selecting between TM and locks on the basis of a cost-benefit analysis using run-time statistics [319]. This uses a syntax `atomic (11) {...}` which can be implemented as either an `atomic` block (when the system is in "transaction mode") or as a lock-based critical section on `11` (when in "mutex mode"). The cost-benefit analysis can consider both the dynamic properties of the critical sections themselves (e.g., whether or not conflicts are frequent) and also the other costs associated with the different implementation techniques (e.g., the run-time overhead that a particular TM implementation imposes). In effect, this means that transactions are used only when they yield benefits. Usui *et al.*'s implementation requires that the programmer ensures that the synchronization used is correct whether it is implemented with locks or with TM. However, one could imagine applying these ideas to select between lock-based and TM-based implementation of `synchronized` blocks with the same semantics.

3.4.3   SPECULATION OVER TM

Futures provide an abstraction for adding concurrency to a sequential program [130]. A program *creates* a future representing a piece of asynchronous work that can proceed in the background, and the program can later *force* the future to retrieve the result of the computation. If the computation has not yet been completed, then this forms a synchronization point and the force operation blocks until the result is ready.

Programming with futures requires care: there is ordinarily no concurrency control between the asynchronous work going on in different futures or between this work and the rest of the program. TM-like techniques have been used to develop forms of *safe future* that avoid these problems [230; 331]. With safe futures, the creation and forcing of futures is treated as an entirely transparent annotation: when compared with the original sequential program, mis-placed annotations may slow the program down, but they will not change its result.

It is not sufficient to simply run each safe future transactionally: first, this would introduce races between the future itself and its continuation. Second, although TM would serialize different futures with respect to one another, it would not force this serial order to be consistent with the original sequential program. The first problem can be tackled by, in effect, using strong isolation while safe futures are active. The second problem can be tackled by specializing commit operations so that they follow the program's original sequential order. Similar techniques are used in implementations of thread-level speculation.

The IPOT model of von Praun *et al.* uses similar ideas to support the addition of implicit parallelism into a sequential program [324]. IPOT provides a `tryasync {...}` construct that identifies a block of code as a possible target for asynchronous execution. Additional annotations can identify patterns of data usage—e.g., variables being used in reductions, or "racy" variables on which conflict detection is unnecessary. A form of `finish` block can be used to delimit points at which all asynchronous work (whether speculative or not) must be complete).

The Grace system provides a safe programming model based on fork-join parallelism [30]. It guarantees that the behavior of a parallel implementation of the program will be consistent with a sequential counterpart in which thread spawns are treated as normal synchronous function calls, and locking and joins become no-ops. This eliminates lock-induced deadlocks, race conditions, atomicity violations, and non-deterministic thread ordering. In return for these guarantees, Grace requires that the program's sequential execution is meaningful, in the sense that it cannot use threads that run indefinitely, or which involve inter-thread communication via condition variables. Grace uses a TM-style implementation based on virtual memory page-level conflict detection and versioning, building on techniques developed by Ding *et al.* for speculative parallelization of sequential code [88].

In all of these models, the possibility to revert to normal, non-speculative execution provides a fall-back to be used in the case of system calls or other work that cannot be made speculatively.

3.5 SUMMARY

In this chapter, we have introduced language constructs based on TM, examined how they can be implemented, how to provide integration between these constructs and the rest of a modern language, and, finally, how TM-based programming has progressed in practice.

Integrating `atomic` blocks in a mature existing programming language is clearly a major undertaking, interactions with other abstractions like locks, condition variables, and volatile data all introduce complexity. To approach this in a methodical manner, definitions like single-lock atomicity (SLA) or transactional sequential consistency (TSC) can be extended to specify the ideal behavior for these constructs, and then practical implementations can be developed by providing this behavior for race-free programs. Building a full implementation that combines these features is clearly a large piece of work, but the design choices and pitfalls are increasingly well understood.

As we have illustrated, `atomic` blocks are not the only way in which TM can be exposed to programmers; it is important to distinguish different possible workloads and the different motivations that exist for building them using transactions. One recurring alternative – going back to some of the earliest work on TM – is simply to expose TM as a programming library to help implement nonblocking data structures or other low-level synchronization primitives. For instance, a 4-word atomic-compare-and-swap operation would drastically simplify the implementation of the double-ended queue from Chapter 2. Compared with general-purpose `atomic` blocks, this kind of use in specialized data structures is a very different proposition: it may place a greater emphasis on performance and scalability, and a lesser emphasis on ease-of-programming and portability (much like specialized algorithms using vector-processing extensions are tailored to the facilities of a given processor family).

CHAPTER 4

Software Transactional Memory

In this chapter, we describe the techniques used in software implementations of transactional memory (STM). We focus on STM systems that operate on conventional processors, deferring until the next chapter architecture support for hybrid software/hardware combinations and the design of new hardware features to accelerate STM.

Shavit and Touitou's paper [285] coined the term "software transactional memory". However, the programming abstraction was rather different than the TM interface from Chapter 2; Shavit and Touitou's design required a transaction to provide, in advance, a vector listing the memory locations that it might access, and to express its proposed memory updates as a function that maps the values seen in these locations to the new values to store back in memory. This approach inspired many early STM implementations, particularly nonblocking designs.

The performance of recent STM systems, particularly those integrated with an optimizing compiler, has reached a level that makes current systems a reasonable vehicle for experimentation and prototyping. However, it is still not clear whether the overhead of STM can be reduced to practical levels without hardware support. STM systems, nevertheless, offer several advantages over HTM:

- Software is more flexible than hardware and permits the implementation of a wider variety of more sophisticated algorithms.

- Software is easier to modify and evolve than hardware.

- STMs can integrate more easily with existing systems and language features, such as garbage collection.

- STMs have fewer intrinsic limitations imposed by fixed-size hardware structures, such as caches.

Given the limited experience that exists implementing and using TM, these considerations suggest that STM will continue to play an important role.

We start this chapter by introducing core implementation techniques that are common across a large number of STM systems (Section 4.1). We describe different ways that the STM's concurrency control metadata is associated with the application's data and ways in which the STM's logs are structured.

Beyond these core techniques, a linear presentation of a rich set of research work inevitably becomes a compromise among competing imperatives. A chronological presentation can capture the large-scale evolution of the area but may obscure smaller-scale interactions by separating related papers. Alternatively, grouping papers by topic raises questions of which are the most important dimensions of the research area and which contributions most clearly define a paper.

The organization of this chapter is one such compromise. We structure the majority of our discussion around a series of detailed case studies of four different kinds of implementation technique. Each of these is illustrative of techniques used in state-of-the-art STM systems, and the selection between them typically represents a fundamental aspect of the design of an STM system.

The first of these case studies (Section 4.2) describes the use of per-object *versioned locks* which combine a lock to control write access to data along with a version number which can be used to let reading transactions detect conflicts. Typical configurations of McRT-STM [7; 274] and Bartok-STM [138] use these techniques. We discuss techniques that were developed to reduce the straight-line overhead of conflict-free transactions—for instance, the use of eager version management and static analyses to optimize the placement of TM operations in `atomic` blocks.

The second case study (Section 4.3) describes the use of a *global clock* along with per-object STM metadata. These techniques can readily support STM systems that provide opacity—that is, STM systems that guarantee that a transaction always sees a consistent view of memory as it runs. The TL2 STM system [83] popularized this technique. We describe the basic design, along with extensions and refinements that both improve the scalability of access to the clock itself and increase the level of concurrency between transactions that synchronize using it.

In our third case study (Section 4.4), we examine STM systems that dispense with per-object STM metadata entirely and *only* use a fixed amount of global state for detecting conflicts. With care, non-conflicting transactions can still run in parallel. Global metadata might seem to pose a problem with respect to scalability. However, on some systems, it is entirely appropriate—e.g., on the Niagara-II CMP, threads on the same chip have access to a common cache, so the metadata is shared by the threads involved. In other cases, simplifications to the STM implementation may compensate for any loss in scaling over the intended workloads. JudoSTM [236], RingSTM [305], and the NOrec STM [77] are examples of systems with global metadata.

Our final case study (Section 4.5) looks at techniques for making *nonblocking* STM systems—as opposed to the lock-based systems of Sections 4.2–4.4. There has been cross-fertilization between blocking and nonblocking STM systems: the most recent nonblocking designs, e.g. Marathe *et al.*'s [208] and Tabba *et al.*'s [313; 314], combine techniques such as those from Sections 4.2–4.4 with fall-back mechanisms to retain strong progress guarantees.

In these case studies, we focus on the techniques used for managing metadata and for detecting conflicts between transactions. As we have illustrated through Chapters 2–3, a full STM system needs to provide more than that. For instance, it may require guarantees about how transactional and non-transactional code interacts, and mechanisms for condition synchronization, for irrevocability, and so on. We discuss these techniques in Section 4.6, looking at related ideas that have been employed in different STM systems.

Section 4.7 describes STM systems that span clusters of machines, rather than operating within a single system. Distributed operation changes the performance characteristics of the underlying system, and if transactions must span separate physical address spaces, then distributed algorithms are needed to support them.

Finally, in Section 4.8, we survey work on testing and validating the correctness of STM systems—for instance, using model-checking to exhaustively test parts of an STM algorithm and using testing frameworks to exercise an STM with "tricky" workloads.

Some Notes on Terminology and Pseudo-Code.   Terminology varies somewhat between papers, and so for consistency, we have tried to use a common set of terms throughout the discussion. A *transaction descriptor* is the per-transaction data structure that keeps track of the state of the transaction; it might be as simple as a single-word status field saying whether or not the transaction has committed, or it might involve additional information such as logs of the transaction's accesses. An *undo-log* holds the values that have been overwritten in an STM using eager version management, and a *redo-log* holds the tentative writes that will be applied during commit in a system using lazy version management. A *read-set* or *write-set* tracks the memory locations that the transaction has read from or written to; it does not necessarily keep track of the data values being manipulated by the transaction. In practice, many systems combine the representation of the undo-log or redo-log with the read-set and write-set.

When writing pseudo-code, we assume that a processor provides sequential consistency (SC). In practice, memory accesses can be re-ordered by an optimising compiler or by an actual processor with a relaxed memory model. The details vary between languages [43; 207] and between processors [9], so care is needed when implementing an algorithm in a particular setting. Guerraoui *et al.* developed techniques to automate the process of introducing memory fences to support STM algorithms [121], and Spear *et al.* developed techniques to optimize the placement of them [304].

4.1   MANAGING STM LOGS AND METADATA

In this section, we introduce basic techniques which are used across most STM systems: maintaining metadata (Section 4.1.1), undo-logs or redo-logs (Section 4.1.2), and managing read-sets and write-sets (Section 4.1.3).

4.1.1   MAINTAINING METADATA

Most STM systems require a mechanism for associating concurrency-control metadata with the locations that the program is accessing. There are two main approaches:

- In an *object-based* STM, metadata is held with each object that the program allocates—this might be an actual object, in a language like C++, or it might be a block of memory returned by `malloc` in C. All fields within the object are associated with the same piece of metadata, and the object's representation in memory is extended to provide space to hold it, for instance, by expanding the object's header. Figure 4.1(a) shows an example, with the three objects each having their own metadata.

Meta-data
Field 1
Field 2
Field 3
...
Field n

Meta-data
Field 1

Meta-data
Field 1
Field 2
Field 3

(a) Object-based organization, with separate metadata added to each object and used to control access to all of the object's fields.

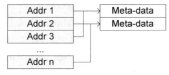

(b) Simple word-based organization, with separate metadata for each word of storage.

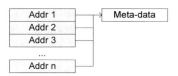

(c) A fixed size metadata table, with a hash function used to map addresses to their corresponding metadata.

Addr 1	→	Meta-data
Addr 2		
Addr 3		
...		
Addr n		

(d) A single, process-wide piece of metadata used for all addresses.

Figure 4.1: STM metadata organization.

- In a *word-based* STM, metadata is associated with individual memory locations, rather than complete objects. Typically, there will be a fixed-size set of metadata locations, and a hash function will map an address onto the location of the metadata value that arbitrates access to it. Figure 4.1(b)–(d) show three examples—first, with the extreme case of a separate piece of metadata for each word in memory, next with a hash function used to map addresses to pieces of metadata, and, finally, the other extreme case of a single piece of process-wide metadata.

The key questions to consider when comparing these approaches are (*i*) the effect of the metadata on the volume of memory used, (*ii*) the effect of accessing the metadata on the performance of the program, (*iii*) the speed of mapping a location to its metadata, and (*iv*) the likelihood of false conflicts between concurrent transactions. Each of these considerations can affect performance, and there are tensions between the different factors. Reducing the volume of metadata can reduce the overhead it adds—both in memory itself and in the caches. However, having a small number of metadata locations can introduce false conflicts between transactions that access the same metadata but different locations [348]. Various "second chance" schemes and alternative validation mechanisms have been developed to detect and reduce the impact of these false conflicts (Section 4.4).

The computational overhead of mapping from an address to a metadata location can be reduced by using a simple mapping function. Several word-based STM systems use modulo arithmetic to map an address onto a slot in a table of metadata values. With object-based STM systems, if the metadata is in the object's header, then it is readily found when accessing fields of the object. Finding the object header can be more complex when using address arithmetic: in such cases, we require a mechanism to map an interior pointer back to the start of the enclosing object—e.g., size-segregated heaps [327], program transformations to keep object header addresses available [81], or tables to map interior pointers to headers [138].

There are three main advantages to placing metadata within an object's header. It can often mean that it lies on the same cache line as the data that it protects, thereby reducing the total number of cache lines accessed by an operation. It can also allow a single metadata access to be made for a series of accesses to different fields of the same object. In addition, object-based designs make the sources of conflicts explicit, and under a programmer's control, rather than depending on the runtime system's placement of data and how this placement interacts with a hash function.

Equally, there are two downsides of using per-object metadata. First, there will be a conflict between concurrent accesses to different fields of the object—or, in many implementations, to different elements of the same array. Avoiding these conflicts can require manual reorganization of data structures. The second downside of placing STM metadata in an object's header is that it can introduce coupling between the STM implementation and the memory management system [83; 162]—this is necessary to avoid problems if the object is deallocated, re-used for some other purpose, and then the old location holding the metadata is accessed by a zombie transaction.

Hybrid metadata organizations may also be used: for instance, using simpler metadata to protect objects that are expected to be read-only and using per-word metadata for large objects or for arrays [261; 265].

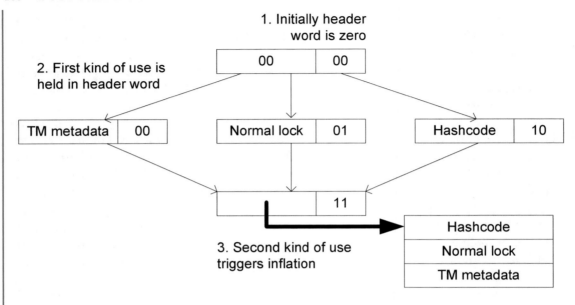

Figure 4.2: Multi-use object header word.

Many language runtime systems already add header information to each object, and techniques have been developed to allow multiple parts of a runtime system to share space rather than reserving multiple separate words in every object [12]—e.g., a hash value might be needed or an ordinary lock to support features such as Java's synchronized blocks. These techniques can be extended to accommodate STM. Figure 4.2 shows the approach taken with the Bartok-STM system [138]. Each object has a single header word, and the low two bits distinguish between different uses. The header word is then *claimed* by the first use that is encountered (e.g., STM, hashing, or a normal lock), and the header word is updated accordingly. If a second use is encountered then the object header is *inflated* to refer to an external structure that contains space for multiple uses. This approach can be used to avoid needing to allocate space for STM metadata on every object in the heap: space is needed only on objects that are actually used during a transaction.

4.1.2   UNDO-LOGS AND REDO-LOGS

STM systems using eager version management require an undo-log of values that will be restored to memory if the transaction is rolled back. STM systems using lazy version management require a redo-log of values that will be written to memory if the transaction commits. In either case, the logs are usually structured as linked lists of chunks so that they can be dynamically extended if necessary depending on the size of a transaction.

With eager version management, the design of the undo log is not usually seen as a performance-critical decision. After all, if the conflict rate is low, the actual values recorded in the log are not required. Nevertheless, a number of aspects of the design require care. First, the granularity at which logging is done can affect the semantics offered by the TM: if the logging granularity is wider than the access granularity, then a roll-back can introduce GLU problems (Section 2.2.3). To avoid this, either the log needs to record the size of the data actually being accessed, or the granularity of conflict detection must be at least as large as the granularity of logging (e.g., using memory protection hardware [5; 28]).

Second, the information recorded in the log entries can affect the ease with which roll back is performed. For instance, in an object-oriented language, there is a question of whether to record entries in terms of an offset relative to an object reference or whether to record entries as a simple address in memory. Similarly, there is the question of whether or not to record the type of the value being overwritten. Recording simple (`addr`,`val`) pairs may reduce the size of the log, but recording an object reference may help integration between the STM and the GC—for instance, allowing the object reference to be updated if the GC relocates the object. Recording additional information, such as types, can increase the size of the log, but it may simplify roll-back (for instance, if the STM needs to perform GC-barrier work when updating reference-typed data during roll-back).

With lazy version management, the design of the redo-log is critical to performance because a transactional read must see the results of an earlier transactional write to the same location. Some early STM systems used direct searching through logs, and so a transaction that writes to n locations may require $O(n)$ operations for each read. This can lead to disastrous performance, even in the absence of contention. There are three main ways to avoid searching:

- An auxiliary look-up structure can provide a mapping from an address being accessed to a previous redo-log entry. This could be a form of hashtable, mapping addresses to positions in the log. Care is needed to keep the number of cache lines accessed during the mapping low, and to provide a "flash clear" operation to avoid iterating through the whole table to clear it (e.g., a version number can be used in entries, or a per-transaction salt combined with the hash function to ensure that the same address hashes to different buckets on successive transactions [138]). Spear *et al.* discuss the performance impact of different approaches [300].

- A summary of the write-set can be maintained (e.g., using a Bloom filter [33]), and the log searched only when reading from a location that might be present in the write-set. However, Felber *et al.* [103] show that the filter can become saturated on some workloads and ultimately degrade performance due to searching on every `ReadTx` operation.

- The STM metadata can provide links to the redo-log entries of the transactions that are currently writing to locations controlled by the metadata. This is particularly simple if the STM system uses eager conflict detection and allows only one transaction at a time to make tentative writes to locations controlled by a given piece of metadata.

The performance of an STM system using lazy version management can be highly dependent on the design of the redo-log, and so when comparing systems, it is important to quantify the impact of this aspect of the design versus the impact of the concurrency-control mechanisms themselves.

4.1.3   READ-SETS AND WRITE-SETS

Most STM systems require a mechanism for a transaction to keep track of the locations that it has read from and written to. With pessimistic concurrency control, this is needed so that the transaction can release any locks that it has acquired. With optimistic concurrency control, it lets the transaction detect conflicts.

STM systems differ in terms of how the read and write-sets are maintained. Using a separate log for each purpose, allows the format of the entries to be customized—e.g., if per-object metadata is used, then read/write-set entries may refer to object addresses, but the undo/redo-log information may refer to individual fields. In addition, using separate logs allows all of the entries of a particular kind to be found easily—e.g., when committing a transaction with eager version management, it is not necessary to skip over undo information which might otherwise be mixed with read/write-set information.

On the other hand, using a single combined log may make it more likely for the current pointer into the log to be kept in a register and may avoid the need to create multiple log entries relating to the same data item.

4.2   LOCK-BASED STM SYSTEMS WITH LOCAL VERSION NUMBERS

Our first case study examines the techniques used in STM systems which combine (*i*) pessimistic concurrency control for writes, using locks which are acquired dynamically by the STM and (*ii*) optimistic concurrency control for reads, implemented by checking per-object version numbers during validation. These are "local" version numbers, in the sense that they are incremented independently on each piece of STM metadata whenever an update is committed to it—as opposed to the systems in the next section, which obtain version numbers from a global clock.

The main algorithmic design choices are between eager and lazy version management, and between acquiring locks when a transaction first accesses a location (encounter-time locking, ETL), or only acquiring them when a transaction commits (commit-time locking, CTL). ETL can support either eager or lazy version management, but it means that conflicts are detected between running transactions, whether or not they commit. Conversely, CTL can only support lazy version management (because the STM cannot update memory directly until locks have been acquired). This allows CTL to support lazy conflict detection.

Dice and Shavit [85] and Saha *et al.* [274] examine the trade offs between these approaches, and Spear *et al.* recently revisited these questions [300] (Section 2.3.2).

In this section, we focus on the combination of ETL and eager version management as a running example. This has been used in many STM systems [5; 7; 138; 273; 274]. These design choices were motivated by performance considerations on traditional multiprocessors and early multicore CMPs: eager version management to avoid re-directing reads through a redo-log and optimistic concurrency for readers to avoid introducing additional contention in the memory system.

In this section, we introduce the main implementation techniques used by this kind of STM system: the design of *versioned locks* for STM metadata (Section 4.2.1), the use of static analyses to optimize the use of STM operations (Section 4.2.2), and the development of techniques to provide opacity in STM systems using local version numbers (Section 4.2.3).

4.2.1   TWO-PHASE LOCKING WITH VERSIONED LOCKS

The STM metadata used in McRT-STM and Bartok-STM is a *versioned lock*. This abstraction combines a mutual-exclusion lock, which is used to arbitrate between concurrent writes, with a version number that is used for conflict detection by readers. If the lock is available, then no transaction has pending writes to the object, and the versioned lock holds the object's current version number. Otherwise, the lock refers to the transaction that currently owns the object.

Compared with the basic TM interface from Chapter 2, McRT-STM and Bartok-STM both use a form of *decomposed* interface in which the `ReadTx`/`WriteTx` operations are broken down into an initial "Open" operation that performs concurrency control and then a separate operation to carry out a data access. For instance a `ReadTx` operation can be expressed as `OpenForRead` operating on the object, followed by an actual data access to one of the object's fields. Similarly, an integer-typed `WriteTx` operation is split into `OpenForUpdate`, followed by `LogForUndoIntTx`, followed by the actual write. This decomposition lets the operations be moved independently during compilation. For instance, if an object-based STM is used, then one `OpenForReadTx` call could be performed before a series of accesses to different fields in the same object.

Before reading from an object, a transaction records the object's version number in its read-set:

```
void OpenForReadTx(TMDesc tx, object obj) {
    tx.readSet.obj = obj;
    tx.readSet.version = GetSTMMetaData(obj);
    tx.readSet ++;
}
```

Before writing to an object, a transaction acquires the object's versioned lock and then adds the object reference and the object's old version number into the write-set. Deadlock must be avoided, whether by timeouts on lock acquires, dynamic detection mechanisms [176], or aborting, releasing other locks, and then waiting for the additional lock that was needed before re-execution [297].

Before writing to a particular field, the system must record the field's old value in the transaction's undo log, so the modification can be rolled back. For example, the runtime system's function to record an overwritten value of type `int` can proceed as follows: (note that the `Magic` class provides low-level memory access, available only from within the language runtime system itself).

```
void LogForUndoIntTx(TMDesc tx, object obj, int offset) {
    tx.undoLog.obj = obj;
    tx.undoLog.offset = offset;
    tx.undoLog.value = Magic.Read(obj, offset);
    tx.undoLog ++;
}
```

Pessimistic locking prevents conflicts on objects that a transaction has written to. However, at commit time, the treatment of reads requires greater care because there is no attempt to prevent read-write conflicts from occurring during a transaction's execution. It is therefore necessary to detect these conflicts via commit-time validation, in order to ensure that the transaction as a whole appears to execute atomically. In pseudo-code, a commit operation proceeds as:

```
bool CommitTx(TMDesc tx) {
  // Check read-set
  foreach (entry e in tx.readSet) {
    if (!ValidateTx(e.obj, e.version)) {
      AbortTx(tx);
      return false;
    }
  }
  // Unlock write-set
  foreach (entry e in tx.writeSet) {
    UnlockObj(e.obj, e.version);
  }
  return true;
}
```

The first loop validates each entry in the read-set in turn, checking whether or not there was a conflict on the object involved. Assuming that this succeeds, the second loop releases all of the write locks that were acquired by the transaction. Releasing a versioned lock increments the version number, and so this will signal a conflict to any concurrent readers of the object. During validation, there are seven cases to consider, based on the values recorded in the transaction's logs, and the current state of the object's versioned lock (Figure 4.3):

- Figure 4.3(a)–(c) show the conflict-free cases. In Figure 4.3(a), the transaction read from the object, and its version number is unchanged at commit time. In Figure 4.3(b), the transaction added the object to its read-set and later added it to its write-set: the two version numbers logged are equal, confirming that there was no intervening conflict. In Figure 4.3(c), the transaction opened the object for writing and then subsequently opened it for reading.

- Figure 4.3(d) shows a simple conflict in which the version number has changed between the time when the transaction opened the object for reading and when the transaction tries to commit.

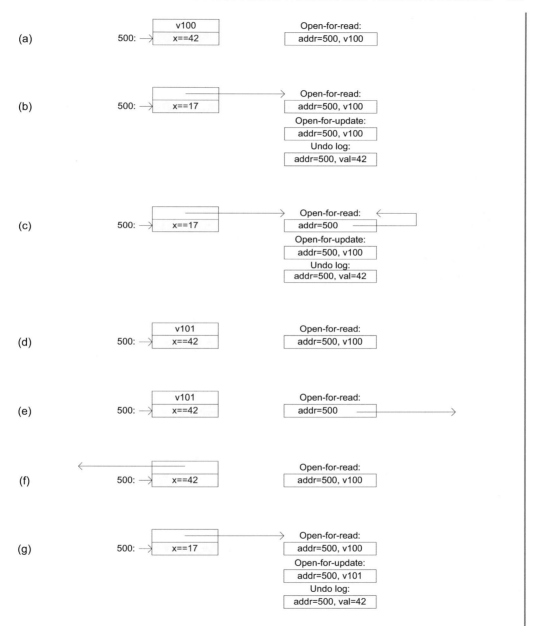

Figure 4.3: Conflict detection using a versioned lock. The left hand side shows an object at address 500, with a single field x, and a versioned lock. The right hand side shows the contents of the transaction descriptor. Arrows to the far left or right denote pointers to another thread's descriptor.

- Figure 4.3(e) shows the case when the object was open for update by another transaction at the point where it was recorded in the read log, and in Figure 4.3(f), the object was open for update by another transaction at the point of the commit operation.

- Finally, Figure 4.3(g) shows the case where one transaction opened the object for reading and later opened the same object for writing, but between these steps, another transaction updated the object. The conflict is detected because the version numbers in the read log and write log do not match.

If validation completes successfully, then a transaction appears to execute atomically at the *start* of its CommitTx operation—even though this is *before* the point at which the outcome of the transaction is known. To see why this is correct, note that for a successful transaction, the subsequent validation work confirms that there have been no conflicts from the point where the transaction initially read the object, until the point where validation occurs. For locations that have been updated, the use of pessimistic concurrency control prevents conflicts until the lock is released. Figure 4.4 illustrates this graphically: the start of CommitTx provides a point at which all of the reads and all of the writes were free from conflicts.

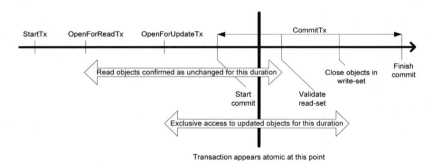

Figure 4.4: Providing atomicity over a set of locations using versioned locks.

Note that opening for read neither checks if the object is open by another transaction nor checks that the transaction's reads form a consistent view of the heap. This design decision is motivated by keeping the fast-paths of conflict-free transactions as simple as possible and minimizing cache invalidations by not forcing readers to modify metadata.

4.2.1.1 Version Number Overflow
If version numbers are large, then the problem of overflow is a theoretical concern rather than a practical one. However, overflow may become a genuine problem if the version number is packed into a 32-bit word along with a lock-bit and, possibly, other bits used by the language implementation.

If overflow is deemed rare, then one approach is to detect when it would occur, then to prevent threads from executing transactions, and to globally re-number the objects so that the version numbers start from 0 again.

Alternatively, given that version numbers are compared only for equality, it is possible to tolerate overflow so long as the "old" use of a given number can never be confused with the "new" use. Bartok-STM exploits this observation: if there are n different version numbers, then it suffices to ensure that every thread validates its current transaction within the time that n version-number increments could occur [138]. This observation avoids the need to re-number all of the objects in the heap.

4.2.2   OPTIMIZING STM USAGE

Decomposing TM operations into separate `OpenFor*` and `LogFor*` steps enables the placement of these low-level operations to be optimized during compilation. Many traditional program transformations apply automatically to STM. For instance, treating the result of an `OpenForReadTx` operation as a form of available expression will allow CSE to remove redundant open-for-read operations. Many STM-related optimizations are possible:

- Eliminating `OpenForReadTx`/`OpenForWriteTx` operations when the target object has already been opened.

- If write access in the STM also grants read access, then an `OpenForReadTx` is redundant if it is dominated by an `OpenForWriteTx` on the same object.

- Programs frequently contain read-modify-write sequences, e.g., `o.a=o.a+1`. These would ordinarily be implemented using an `OpenForReadTx` followed by an `OpenForWriteTx`. The `OpenForReadTx` operation can be avoided by moving the `OpenForWriteTx` operation earlier in the code.

- Analyses to identify immutable objects, or newly-allocated thread-local objects, can allow `OpenForRead*` operations to be removed. A simple data-flow analysis can identify some common cases. In Java, `final` fields are immutable in most code, and built-in abstractions such as `java.lang.String` objects are known to be immutable.

- Additional redundancies may be exposed by moving `OpenFor*` operations to a method's caller.

- The buffer overflow checks in the logging operations in a transaction can sometimes be optimized into a single test, which checks whether there is sufficient space for all of the data logged in the transaction.

These optimizations can reduce the number of log entries by 40–60% on a variety of benchmarks, with a roughly similar improvement in a transaction's execution time.

Performing these transformations requires care. In the compiler's internal intermediate code, the STM operations must be explicitly connected to a specific transaction so a compiler does not optimize across transaction boundaries and to ensure that the sequencing of operations is respected. This can be done by encoding the relationships as data-dependencies between the operations—so-called pseudo-variables or proof-variables [221] that exist within the compiler representation, but not the generated code.

For instance, an assignment `obj.f1=42` would be translated into intermediate code of the form:

```
tx = GetTxDescriptor();
<p1> = OpenForWriteTx(tx, obj);
<p2> = LogForUndoIntTx(tx, obj, 4) <p1>;
obj.f1 = 42 <p2>;
```

The first operation, `GetTxDescriptor`, fetches the current transaction descriptor from thread-local storage. This can be re-used across multiple STM operations, rather than being fetched each time. `OpenForWriteTx` performs concurrency control to allow the transaction to access `obj`. The `LogForUndoIntTx` operation logs the old value of the first field of `obj` (assuming this is at offset 4 from the object header). Finally, the write itself is performed. The variables p1 and p2 are both proof-variables, and the notation identifies where they are needed as inputs (on the right of an operation), or produced as results (on the left). The dependencies via tx, p1, and p2 prevent incorrect re-orderings—for instance, they ensure that the logging is performed before the write.

The compiler must be aware that the implementation of `GetTxDescriptor` is dependent on the current transaction. For instance, the transaction descriptor can be cached in a register within a transaction, but not across the start or end of a transaction.

Wu *et al.* describe additional optimizations applicable to an STM using lazy version management [336]. Thread-local data can be read and written directly, rather than being written via a redo-log (this requires checkpointing any thread-local data that is live at the start of the transaction, and care on indirect writes, to detect whether they access thread-local data, or shared data).

Additional techniques can be used dynamically to eliminate redundant log entries that are not removed by compile-time optimizations [138]—e.g., dynamically tracking which objects are allocated inside a transaction (and eliminating undo-log entries for any updates to them). If dynamic techniques are used then their expense must be traded off against the reduction in log-space usage that they achieve.

4.2.3   PROVIDING OPACITY

McRT-STM and Bartok-STM typically use *invisible reads*, meaning that the presence of a reading transaction is not visible to concurrent transactions which might try to commit updates to the objects being read. The basic versions of these STMs allow a transaction to experience a conflict and to continue executing as a zombie; consequently, they do not provide opacity. As we discussed in Section 2.2.2, one way to obtain opacity is to add incremental validation. However, incremental validation is likely to be slow because a transaction reading n locations will perform $O(n^2)$ individual validation checks in its execution.

To avoid incremental validation, an STM system could dispense with invisible reading and maintain explicit visible read-sets. Some STM systems offer this as an option (e.g., some configurations of McRT-STM, and also RSTM, Section 4.5.2), but if applied to all transactions on a conventional system, visible reading introduces contention between readers in the memory system. Consequently, visible reading has often been found to be costly [274].

An alternative option is to use a global clock. This approach is taken by many STM systems, and we return to it in detail in the next case study (Section 4.3). However, instead of using a global clock, there are several techniques which can be used to mitigate the cost of incremental validation in a system such as McRT-STM or Bartok-STM:

4.2.3.1 Global Commit Counter

Spear *et al.* observed that validating a transaction (TA) is necessary only if another transaction (TB) has updated memory since the most recent time that TA was validated [302]. To exploit this with lazy version management, they introduced a *global commit counter* and have each thread re-validate only if the counter has been incremented since its previous validation. This approach is simple and is often effective. A limitation of it is that *every* transaction will be revalidated, even if there is no conflict involved.

4.2.3.2 Semi-Visible Reads

Lev *et al.* [188; 191] introduced a form of *semi-visible* reading. This allows a writer to detect that it is accessing data being used by reading transactions but it does not let the writer detect the identity of the accessors.

To maintain this information, Ellen *et al.*'s Scalable Non-Zero Indicators (SNZI [97]) can be used. A SNZI counter supports `Inc` and `Dec` operations but only an `IsZero` query, rather than access to the underlying value. A scalable implementation is based on a tree of individual non-zero-indicators. `IsZero` is always invoked on the root but `Inc` and `Dec` can be invoked on any tree node (so long as a given thread is consistent in its choice of node). The interior nodes propagate `Inc` and `Dec` toward the root but only if their own local count changes to/from zero; consequently, the tree filters out contention under high rates of increments and decrements.

A SNZI counter is associated with each piece of STM metadata, and `Inc` is invoked when a transaction starts to read from a location, and `Dec` invoked when it commits or aborts. A writer invokes `IsZero` to detect whether or not readers may be present. If a writer sees that readers may be present, then it increments a global commit counter. Readers use changes in the global commit counter to trigger re-validation. This can result in substantially less re-validation work because a writer needs to increment a global counter only when there is a possible conflict, rather than for every commit.

4.2.4   DISCUSSION

In this section, we have introduced the techniques used in STM systems such as McRT-STM and Bartok-STM which combine pessimistic concurrency control for updates with optimistic concurrency control for reading.

The underlying STM algorithms provide fairly weak semantics: (*i*) they do not provide opacity during transactions, (*ii*) they do not provide conflict detection between transactional and non-transactional accesses to memory locations, and (*iii*) they do not all support idioms such as the privatization example from Section 2.2.3.6. If the programming model requires these properties, then they must be built over the core STM interface—for instance, by sandboxing faults that occur and by expanding non-transactional accesses to interact with the STM.

Concurrent with McRT-STM and Bartok-STM, Ennals investigated a library-based STM, which was also based on versioned locks [99]. Rather than using an undo log, Ennals' design uses lazy version management and has transactions take a private snapshot of entire objects that they are working on. A transaction then operates on the snapshot and, if it commits, writes the snapshot back to shared memory. Ennals' design enables lock revocation, provided that the transaction owning an object has not yet started committing updates.

Hindman and Grossman's AtomicJava STM system [154] provides `atomic` sections based on extending each object with a field that identifies a thread that currently owns the object. Source-to-source translation is used to add lock acquire/release operations. Pessimistic concurrency control is used for reads as well as for writes, and thread-private undo logs are used for roll-back. Unlike most STM systems, locks are transferred between threads by having each thread poll for requests to release a lock. This approach allows a lock to remain associated with a thread across a series of operations without being released and re-acquired.

4.3   LOCK-BASED STM SYSTEMS WITH A GLOBAL CLOCK

The next series of STM algorithms that we study are based on the use of a *global clock*. This clock is incremented on a process-wide basis (unlike the version numbers used by McRT-STM and Bartok-STM which are incremented separately on a per-object basis).

To illustrate this idea, we focus on the TL2 STM system [83]. Semantically, the key distinction between TL2 and STM systems such as McRT-STM and Bartok-STM is that TL2 provides *opacity* without requiring incremental validation. (TL2 itself developed from an earlier "transactional locking" (TL) STM system that was used to explore various combinations of locking mechanisms and version-number management [84; 85].)

We introduce the original TL2 STM algorithm in Section 4.3.1. Section 4.3.2 discusses *timebase extension*, a technique to provide greater concurrency between transactions than the basic algorithm provides. On some hardware platforms, contention on the global clock may be a worry. However, as we show in Section 4.3.3, many techniques have been developed either to reduce the number of times that a thread needs to update the global clock, or to provide alternative, more scalable, implementations of the clock.

4.3.1 PROVIDING OPACITY USING A GLOBAL CLOCK

Like the STM systems in Section 4.2, TL2 uses versioned locks, which either hold a timestamp value, or identify a single transaction that currently owns the associated data. TL2 uses *lazy version management*, so tentative updates are built up in a redo-log, and it uses *commit-time locking*, so locks are acquired only as part of a transaction's commit function. In addition to the versioned locks, TL2 employs a global clock—this can be a simple 64-bit shared counter, which is incremented each time a transaction commits.

Each transaction begins by reading the global clock at the point when it starts. This timestamp is known as its *read version* (RV). In addition to RV, each transaction is assigned a unique *write version* (WV) from the global clock as part of its commit operation. This is used to define the transaction's position in the serial order. Each object's version number records the WV of the transaction, which most recently committed an update to it.

A transactional read proceeds as follows:

```
int ReadTx(TMDesc tx, object obj, int offset) {
  if (&(obj[offset]) in tx.redoLog) {
    // Tx previously wrote to the location.
    // Return value from redo-log
    result = tx.redoLog[&obj[offset]];
  } else {
    // Tx has not written to the location.
    // Add to read-set and read from memory.
    v1 = obj.timestamp;
    result = obj[offset];
    v2 = obj.timestamp;
    if (v1.lockBit ||
        v1 != v2 ||
        v1 > tx.RV) {
      AbortTx(tx);
    }
    // Add to read-set
    tx.readSet.obj = obj;
    tx.readSet ++;
  }
  return result;
}
```

Since TL2 uses lazy version management, the read operation must first check whether or not the location is already in the transaction's redo-log. If it is in the log, then `ReadTx` returns the value from the log, ensuring that the transaction sees its own earlier write. If the value is not in the log, then `ReadTx` reads the object's version number, reads the value from memory, and then re-reads the version number. This series of reads ensures that a consistent snapshot is taken of the two locations: if `v1.lockBit` is set then the object was locked by a concurrent transaction's commit operation, and if `v1!=v2`, then the version was changed by a concurrent transaction while the value was being read from memory.

The key to providing opacity is that `ReadTx` also checks that the version seen is no greater than `tx.RV`. This check ensures that the object has not been updated by any other transaction since RV was recorded. Consequently, RV provides a timestamp at which *all* of the locations read by a transaction represent a consistent view of memory: commits before RV will have been seen, and a read will abort if it sees a commit after RV.

`WriteTx` is more straightforward: if this is the first write to a location then it adds an entry to the redo-log; otherwise, it updates the current log entry.

In pseudo-code, a commit operation proceeds as:

```
bool CommitTx(TMDesc tx) {
  // Lock write-set
  foreach (entry e in tx.writeSet) {
    if (!Lock(e.obj)) {
      goto fail;
    }
  }
  // Acquire unique WV
  tx.WV = FetchAndAdd(&globalClock);
  // Validate read-set
  foreach (entry e in tx.readSet) {
    if (e.obj.version > tx.RV) {
      goto fail;
    }
  }
  // Write back updates
  foreach (entry e in tx.redoLog) {
    e.obj[e.offset] = e.value;
  }
  // Unlock write-set, setting version number
  foreach (entry e in tx.writeSet) {
    UnlockObj(e.obj, tx.WV);
  }
  return true;

fail:
  // Release any locks acquired
  ...
  return false;
}
```

Figure 4.5 shows this diagrammatically. While the transaction runs, the point at which RV is taken represents the instant at which the transaction's atomic snapshot of memory is valid; the transaction aborts if it reads any object whose version number is newer than RV. The commit operation is bracketed by acquiring and releasing write locks. This prevents conflicting operations on objects in its write-set (as shown by the lower large arrow). The commit operation's validation loop checks that none of the objects in the transaction's read-set were updated after RV. This checks whether there were any conflicting operations on them after RV but before WV (as indicated by the upper large arrow in the figure). Consequently, the transaction can appear to occur atomically at the point where WV was taken. After successful validation, the transaction writes back its updates and releases the write locks.

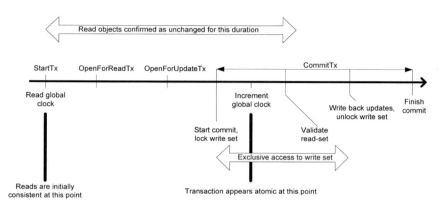

Figure 4.5: Providing atomicity over a set of locations using a global clock.

This basic form of TL2 can be extended in many ways:

Read-only Transactions.   With TL2, a read-only transaction does not actually require *any* logging, beyond a record of RV when it starts. To see why this is possible, notice that with an empty write-set, the CommitTx operation reduces to (*i*) allocating WV and (*ii*) re-validating the read-set. The WV value itself is not used (since there are no updated objects to write the value back into), and removing the allocation of WV can reduce contention on the global clock. The re-validation is not needed because, without any writes, the point at which RV was taken can form the linearization point for the complete transaction.

Adjacent Transactions by the Same Thread.   If CommitTx determines that WV==RV+1, then there cannot have been any concurrent transactions serialized after RV but before the current transaction. Consequently, the re-validation loop is not needed because there cannot have been any conflicts.

Hierarchical Locking.   Felber *et al.*'s TinySTM system combines a TL2-style approach with a form of hierarchical locking [103]. An additional, shared, array is used with slots in the array covering regions of the process's address space. Slots in the array each hold counters of the number of commits to locations in that region.

A transaction maintains read/write masks with one bit per slot, indicating the areas of memory that it has accessed. In addition, upon its first read from a slot, it records the associated counter's current value. If the counter has not been incremented by the time that the transaction attempts to commit, then it is not necessary to validate accesses to that region of memory. This can streamline validation work.

Felber *et al.* also investigate the sensitivity of performance to the sizing of the shared table and to the way that addresses are mapped to normal metadata entries.

Eager Version Management.   Felber *et al.* [103] and Wang *et al.* [327] describe variants of a TL2-style approach which can use eager version management. To do this, encounter-time locking must be used for writes. In addition, *aborted* transactions must update the version number for a location when releasing its lock—if this were not done, then a concurrent call to ReadTx might see v1 before the aborted transaction, and v2 after the aborted transaction, and not detect that the ReadTx call has seen a tentative write.

These version number updates can be problematic under high contention: readers can be forced to abort because of the version-number change, without actually seeing the aborted transaction's write. To address this problem, Felber *et al.* added an *incarnation number* to each piece of STM metadata. An aborted transaction increments the incarnation number, but it does not change the version number. The ReadTx operation checks that the incarnation numbers match in v1 and v2, and that the read has seen a consistent snapshot of the location and its version number.

Version Number Overflow.   As with per-object version numbers, an implementation using a global clock must accommodate overflow. As before, if 64 bits are available then this is essentially a theoretical concern. Felber *et al.* describe a technique for handling overflow in the clock [103]: a transaction detects that a clock has overflowed when it is about to increment it from the maximum possible value. The transaction aborts itself, waits for other transactions to complete, and then resets all version numbers. This aborts some non-conflicting transactions unnecessarily, but for an n-bit clock, it happens at most once every 2^n transactions.

Memory Fences.   A naïve implementation of a TL2-style ReadTx operation may require read-before-read memory fences to ensure that the timestamp seen in v1 is read before the value, and that the timestamp in v2 is read after. The overhead of these fences can be substantial on processors such as the POWER4. Spear *et al.* investigate static analyses for reducing the number of fences that are needed [304].

In the context of TL2, the idea is to decompose reads and writes into pre-validation, fence, access, fence, and post-validation stages, and then to allow re-ordering of these stages with respect to other code, so long as the relative ordering of these stages for independent accesses is retained, and so long as the post-validation stage has been completed before any potentially unsafe use of the value is made (e.g., a computed memory access, or an indirect branch). A fence may be removed if it is adjacent to another fence.

For instance, if a transaction executes a series of K memory accesses, then these analyses allow the set of initial version number reads to be moved into a batch, then a single fence executed, then the actual K accesses made, and, finally, a second fence followed by the second batch of version number reads. Consequently, $2K$ fences are replaced by 2.

4.3.2   TIMEBASE EXTENSION

TL2's use of a global clock guarantees that a transaction sees a consistent view of memory as it runs. However, it can exhibit false positives in which a transaction is aborted despite seeing a consistent view of memory. Consider the following example:

```
// Initially: global clock = 0

// Thread 1                    // Thread 2
StartTx()
                              StartTx()
                              WriteTx(&x,42);
                              CommitTx();

r = ReadTx(&x);
...
```

In this case, both threads use RV=0. Thread 2 then runs its transaction completely, so it receives WV=1, increments the global clock to 1, and sets the version number for x to be 1. Consequently, Thread 1 sees this version number and aborts because it is later than its RV.

The problem here is that there is not really anything wrong with Thread 1's transaction; its view of memory is consistent with a serial order after Thread 2's transaction, but the fact it saw RV=0 forces it to abort unnecessarily.

Riegel *et al.* introduced a mechanism to allow transactions to recover from this kind of false conflict, while still detecting when a genuine lack of consistency occurs [264]. This mechanism was applied to a TL2-style STM by Wang *et al.* [327], Zhang *et al.* [342] and Spear *et al.* [300]. Instead of having ReadTx abort if it reads an object version later than RV, the read operation fetches a new RV' value from the global clock, and then validates its existing read-set according to the original RV. If this validation succeeds, then RV' can replace RV.

In the previous example, Thread 1 would set RV'=1, validate its (empty) read-set, and then continue with RV=1. Conversely, if Thread 1 had read from x *before* Thread 2's update, then the validation would fail, and Thread 1's transaction would need to be aborted.

Sadly, timebase extension means that the racy publication idiom is no longer supported. This is because, referring to the execution example in Section 2.2.3.6, the validation check succeeds for the data x and the flag x_published even though there was an intervening non-transactional write to x. Whether or not this is important depends on the programming model that the STM wishes to support; it illustrates the kind of trade-off that exists between the semantics provided and the implementation techniques that are correct.

4.3.3   CLOCK CONTENTION VS FALSE CONFLICT TRADEOFFS

On some systems, the fact that every read-write transaction needs to update the global clock might be a performance worry. Several techniques have been developed which allow threads to avoid incrementing the global clock on every transaction or which replace a simple integer clock with a more distributed implementation. In many cases, these represent a tradeoff between reduced contention on the global clock and a loss of concurrency between transactions.

Timestamp Re-Use Across Threads.   Dice *et al.* observe that introducing a thread ID into the timestamps permits some atomic compare-and-swap (CAS) operations on the global clock to be removed [83]. Each object's STM metadata is extended to include the ID of the thread that most recently committed an update to the object. To obtain its new write version (WV), a thread compares the current global clock with the write version from its own most recent transaction. If the numbers differ, then the thread uses the global clock without incrementing it. If they do not differ, then the thread increments the global clock as normal. The combination of this ID along with WV provides a unique ID for the transaction. In the best case, this allows every thread to use the *same* version number before incrementing the global clock.

The `ReadTx` function must be modified to detect a conflict on a read equal to RV. This can introduce false conflicts, since it is not possible to distinguish a valid read from an earlier transaction using a given version number, versus an invalid read from a later transaction with the same number.

Non-Unique Timestamps.   Zhang *et al.* observe that, even without introducing thread IDs, there are some situations where a timestamp can be reused [342]. In particular, if a set of transactions commits concurrently and are all validated against one another at commit-time, then they can share a timestamp. In effect, the validation ensures that there are no conflicts within the set, and a common timestamp is sufficient to signal conflicts to transactions outside the set. The paper shows how to exploit this observation by providing several variants of the `CommitTx` function.

Lev *et al.* describe this as a "pass on failure" strategy: an atomic CAS does not need to be reexecuted if it fails when incrementing the global clock [188].

Avoiding False Updates.   Zhang *et al.* observe that the original TL2 `CommitTx` function increments the global clock even for transactions that fail validation [342]. In the paper, they examine ways to exploit this observation, in different versions of the `CommitTx` function, and associated changes to `ReadTx`.

Deferred Clock Update.   Lev *et al.* describe a variant of TL2 termed GV5. With GV5, a transaction does not write back the incremented value to the global clock, but it still uses the new value when updating the STM metadata [188]. The version numbers in metadata may therefore have values greater than that of the clock. Readers that see such values increment the counter to (at least) the version number that they see. This can cause unnecessary aborts—e.g., one thread repeatedly accessing the same location will abort on every other attempt.

Lev *et al.* also describe a hybrid scheme, GV6, which uses "pass on failure" with probability 1/32, and uses GV5 the rest of the time. They observe that this reduces unnecessary aborts by advancing the counter periodically while still reducing the frequency of updating the counter.

Thread-Local Timestamps.   Avni and Shavit introduced a technique to allow thread-local clocks to be used in place of a common global clock [24]. The STM metadata is replaced by (`thread,time`) pairs, where `time` is a value from a per-thread clock, incremented on every successful commit. In addition to its own clock, each thread has an array of remote clock values for the other threads, representing the latest times that the current thread has seen from each other thread.

In effect, this local array represents a transaction's RV in a form of vector clock. If a transaction running on thread `t1` reads from an object with version `(t2,v)`, then this is treated as being earlier than RV if `v` is less than or equal to `t2`'s value in `t1`'s array. Unlike GV5, this avoids self-aborts because one thread always sees the most up-to-date value for its own clock.

If a thread reads from an object which contains a timestamp later than its value for the writing thread, then it aborts its own transaction and refreshes its value from the other thread's clock. These aborts can introduce false conflicts, and so this technique represents a tradeoff between the work lost from false conflicts, versus the savings made by the reduction in communication between non-conflicting threads.

4.3.4   ALTERNATIVE GLOBAL CLOCK ALGORITHMS

Other STM systems have used a notion of global time. Riegel *et al.* describe a multi-version STM system that provides snapshot isolation [263] and linearizable transactions [262]. Each transaction maintains a timestamp window during which its current snapshot is known to be valid. For instance, after reading the current global clock into RV, this range is initially $[RV, \infty]$. The implementation retains multiple versions of each object, valid over different timestamp windows, making it more likely that a version is available which falls within the transactional timestamp window. If sufficient versions are available, this guarantees that read-only transactions can always commit. As objects are accessed, the timestamp window is reduced by intersection with the timestamps during which the object's data is valid. When accessing the current version of an object, the end of that version's validity is approximated by the current timestamp (since the point at which the version will become invalid is not yet known). Re-validation allows the timestamp window to be extended, so long as there has not been a conflicting update. To provide linearizability, the transaction's timestamp window must extend over its commit timestamp.

SwissTM [93] combines a TL2-style global clock, with mixed conflict detection: write/write conflicts are detected eagerly, and read/write conflicts are detected only at commit time. To accommodate this, Dragojević *et al.* use STM metadata that holds both a version number (for use in validation) and a pointer to an owning transaction descriptor (to permit eager write/write conflict detection). As Spear *et al.* discuss [302], this can allow transactions with read/write conflicts to proceed in the hope that the reader commits before the writer.

4.4   LOCK-BASED STM SYSTEMS WITH GLOBAL METADATA

In this section, we look at techniques for STM systems which use global metadata; that is, the *only* shared metadata are global structures, with no individual locks or version numbers associated

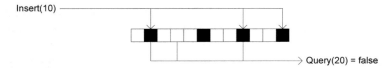

Figure 4.6: A Bloom filter, using 3 hash functions. 10 is inserted by setting the 3 bits indicated. A query of 20 returns false because not all 3 of its bits are set.

with objects in the program's heap. The attraction of these systems is that they can avoid the space overhead and cache pressure of using fine-grained metadata, and they can reduce the number of atomic operations involved in running and committing a transaction.

There are two main challenges in designing an STM system with only global metadata:

First, since the metadata is shared among all of the threads, the design should access it sparingly to reduce the amount of memory system contention. The systems we describe here typically make *per-transaction* updates to the metadata, rather than *per-access* updates. In addition, the JudoSTM [236] and NOrec [77] systems (Section 4.4.2) allow read-only transactions to execute without needing to make any updates to the metadata.

Second, for all but the lowest levels of contention, the STM system needs a way to detect conflicts in terms of the actual locations that a transaction accesses, rather than executing transactions serially. We look at two techniques. The first (Section 4.4.1) is to use Bloom filter summaries of a transaction's read-set and write-set. These can be compared between transactions to determine whether or not the transactions might conflict. The second approach (Section 4.4.2) is to use value-based validation: recording the actual values seen by a transaction and checking these for conflicting updates during commit.

4.4.1   BLOOM FILTER CONFLICT DETECTION

Bloom filters provide a conservative summary of a set's contents [33]. They support constant-time operations to insert and query items, and to test whether or not two sets have a non-empty intersection. Queries on a Bloom filter are conservative, in the sense that *query*(x) may return true, even though x is not actually present in the set. Conversely, if *query*(x) returns false, then the item x is guaranteed to be absent from the set. A Bloom filter operates by using a set of hash functions to map a value x into a set of indices in a bit-vector. An item is inserted by setting the bits at these indices to 1. A query tests all of the indices, and it returns true if they are all set. Figure 4.6 illustrates this; we return to hardware mechanisms for managing Bloom filters in Section 5.3.2.

RingSTM introduced the idea of Bloom filter-based conflict detection in a lazy-versioning STM system [305]. This avoids using any per-location TM metadata in the heap, and it allows a transaction to maintain its read/write-sets by using a pair of thread-local Bloom filters.

RingSTM uses a time-ordered shared list of transaction commit records. Each record holds the transaction's logical timestamp (assigned as part of the commit process), a Bloom filter for its write-set, a current status (WRITING, or COMPLETE), and a priority field. A WRITING transaction is one which has been assigned a position in the serial order but not yet finished writing back its updates. A COMPLETE transaction has finished its writeback phase.

A transaction starts by recording the logical timestamp of the oldest transaction in the list that is WRITING. As a transaction runs, it maintains private Bloom filters for its own read-set and write-set, and a redo-log of its proposed updates. To provide opacity, if desired, a transaction intersects its read-set filter against the write-sets of other transactions when the shared list is modified. (In effect, responding to modifications as with a global-commit-counter algorithm.) To commit, a transaction validates its reads for a final time, and then it uses an atomic CAS to claim a new entry in the shared list. This establishes the transaction in the serial order. It then completes its writes and updates its status to COMPLETE.

The *priority* bit is used to avoid repeated re-execution in the case of conflicts: a thread commits a dummy transaction with the high-priority bit set, and this disables non-high-priority transactions from committing until the thread finishes the work it intended.

Rather than using a physical shared list of commit records, RingSTM organizes the list within a fixed-size ring. Ring entries must be retained until they are COMPLETE. A transaction must abort if its start time becomes older than the oldest ring entry; if this happens, then it will be unable to obtain write-sets of the intervening transactions, and so it cannot detect conflicts reliably.

Mehrara *et al.* [218] describe an STMlite system used for conflict detection between speculatively-executed loop iterations. As with RingSTM, threads maintain private Bloom filters summarizing their accesses. However, commit is managed by a central transactional commit manager (TCM) which uses the signatures to determine which transactions to allow to commit. Once notified of the outcome from the TCM, a thread proceeds to write back the values. No synchronization is needed on these writes because the TCM has already verified that transactions are updating disjoint sets of data.

The InvalSTM system uses a form of commit-time *invalidation* based on Bloom filters [112]. Each transaction descriptor contains a valid flag, read/write Bloom filter summaries, and a redo log. A transaction starts with its valid flag set to true, and, if opacity is required, the transaction checks that the flag remains set after each of its memory reads. For a transaction to commit, it checks its own write Bloom filter against the read/write Bloom filters of each of the other active transactions in the system. If a conflict is detected, then a contention manager decides which of the conflicting transactions should be stalled or should be aborted (by clearing its valid flag). Commits are serialized on a global lock, and per-transaction-descriptor locks are used to protect the individual read/write sets.

Gottschlich *et al.* argue that InvalSTM-style invalidation can have important advantages for high contention workloads—e.g., if a writing transaction TA conflicts with a set of reading transactions TB...TZ, then it is possible for the contention manager to delay TA or to abort it, rather than requiring that all of the readers are aborted. They also showed that InvalSTM-style invalidation can be particularly effective for memory-intensive read-only transactions: opacity is provided by the checks on the `valid` flag, and no commit-time conflict detection is required because the write Bloom filter is empty.

4.4.2   VALUE-BASED VALIDATION

The second technique for conflict detection using simple global metadata is to use a form of *value-based* validation, in which a transaction determines whether or not it has experienced a conflict based on a log of the actual values that it has read from memory, rather than a log of version numbers. However, a simple commit-time value-based check is insufficient. Consider the following problem:

1. Location X's initial value is "A".

2. Transaction T_1 reads this value and records it in its read set.

3. Transaction T_2 changes X's value to "B" and commits.

4. Transaction T_1 reads X again, but this time sees the value "B".

5. Transaction T_3 changes X's value back to "A" and commits.

6. Transaction T_1 commits and validates its read-set. Since X's value has returned to "A", the transaction will pass validation, even though it read two inconsistent values for X.

JudoSTM.   The JudoSTM system [236] ensures that a transaction's reads always see values from its own log (so that T_1 would see the same value for X if it reads from it more than once), and it ensures that transactions serialize their validation and commit operations using locks (so that there cannot be any concurrent changes to X while T_1's accesses are being checked).

JudoSTM supports two kinds of commit: in coarse-grained mode, it uses a single versioned lock to protect validation and commit operations. Writers hold the lock, and validate and commit serially before incrementing the single version number. Readers commit without holding the lock, but they check that the version number is unchanged before and after their work. JudoSTM supports an additional fine-grained mode, in which a hash function maps individual locations to separate locks.

JudoSTM is also notable in using binary translation to add the instrumentation necessary for using TM in C/C++. The translator generates specialized versions of the TM operations rather than using general-purpose logs—e.g., it can specialize the validation code according to the number of locations being accessed.

TML and NOrec.   Dalessandro *et al.* describe a pair of STM systems which, as with JudoSTM, operate using global metadata [77; 307]. These are designed to provide low-cost fast-path operations, livelock freedom between sets of transactions, and commit-time conflict detection.

The first system, TML, uses a single global versioned lock. This is used to directly serialize writing transactions, being held while the transaction runs, and thereby allowing the transaction to execute without maintaining logs of any kind (assuming that user-initiated roll-back is not supported). Read-only transactions record the current version number when they start, and they re-check it after every read in order to maintain opacity. This approach is effective in workloads where reads are the common case. Of course, using a single lock without logging means that conflict detection is extremely conservative: any writer conflicts with any other concurrent transaction.

The second system, NOrec, also uses a single global versioned lock, but unlike TML, it acquires the lock only when updating memory during commit. With NOrec, a transaction maintains a read log of (addr,val) pairs and a snapshot timestamp (RV) taken from the global lock's version number at the point where the transaction began. Writes are made directly into the transaction's redo-log, with a hashing scheme used to avoid searching. Reads are a little more involved:

```
int ReadTx(TMDesc tx, int *addr) {
  // Check for earlier write:
  if (tx.writeSet.contains(addr))
    return tx.writeSet[addr]
  // Read consistent (addr,val) pair:
  val = *addr;
  while (tx.RV != global_clock) {
    tx.RV = global_clock;
    ValidateTx(tx);
    val = *addr
  }
  // Update read-set:
  tx.readSet.append(addr, val)
  return val
}
```

The while loop ensures that the transaction's RV still matches the global clock. If it does not, then the transaction is re-validated: the ValidateTx function waits until the global lock is not held and then uses the (addr,val) pairs in the read log to check whether or not the transaction is still valid. This provides opacity, and ensures that a series of reads from the same location will see the same value (without needing to explicitly search a read-log on every access).

Read-only transactions need no further work to commit since the ReadTx function has already ensured that they were consistent at the point of their last read. To commit a transaction with writes, the global lock's version number is first sampled, the transaction re-validated, and then the global lock is acquired (if the version number has changed in the meantime then the commit operation is restarted). Assuming validation succeeds, then the transaction writes back its log and releases the lock, incrementing the version number. Note how the use of a versioned lock enables the validation phase to execute *without* needing to acquire the lock.

4.4.2.1 Hybrid Use of Value-Based Validation

Value-based validation has also been employed as a "second chance" scheme in the word-based STM of Harris and Fraser (Section 4.5) where a transaction which fails version-number-based validation can refresh its expected version numbers by performing a value-based check [135]. Ding *et al.* used value-based validation in their work on speculative parallelization using process-level virtual memory [88], and Berger *et al.* used value-based validation as part of the Grace system for multithreaded programming in C/C++ [30].

4.5   NONBLOCKING STM SYSTEMS

The final set of STM systems which we consider are those in which the individual STM operations provide *nonblocking* progress guarantees such as lock-freedom or obstruction-freedom (Section 2.3.1). The first STM systems were developed to help implement nonblocking data structures, and so naturally, they are needed to provide nonblocking progress guarantees. In this section, we discuss some of these early systems, along with more recent STM systems which combine nonblocking progress with the kinds of implementation technique from the case studies of Sections 4.2–4.4.

The challenge in building a nonblocking STM system, as with other nonblocking algorithms, is ensuring that the memory accesses made by a transaction all appear to take place atomically, even though the implementation is performing a series of single-word atomic operations. Lock-based STM systems can accomplish this by mutual exclusion: if thread T2 is committing a series of updates, then it locks all of the locations involved, and the locking protocol prevents other threads from accessing those locations until T2 is finished. With a nonblocking STM, it is impossible for T2 to prevent T1 from using the memory locations that are involved in T2's transaction. Instead, T2 must ensure that, no matter how far it has reached through its own commit operation, (*i*) T1 will either see all of T2's updates, or none of them, and (*ii*) if T1 attempts its own, conflicting, memory accesses, then the conflict between T1 and T2 will be detected and resolved.

In this section, we introduce the key ideas used by DSTM, the first object-based nonblocking STM system (Section 4.5.1). We then introduce the broader design space for this kind of STM, which add one or more levels of indirection to each object's representation (Section 4.5.2). Finally, we describe techniques for building nonblocking STM systems without introducing indirection (Section 4.5.3).

4.5.1   PER-OBJECT INDIRECTION

Herlihy, Luchangco, Moir, and Scherer's paper [146] was the first published work to describe a dynamic STM (DSTM) system that did not require a programmer to specify, in advance, the locations that a transaction is going to access (unlike Shavit and Touitou's original STM [285]). DSTM introduced the use of an explicit *contention management* module in STM systems.

DSTM was implemented as a Java library, for use on an unmodified JVM. The programming model used explicit transactions, rather than `atomic` blocks, and required that transactionally-accessed objects were wrapped by instances of a `TMObject` class which provides transactional semantics. These objects must be explicitly opened before being read or written in a transaction:

```
class TMObject {
   // Construct new TM object, wrapping obj
   TMObject(Object obj);
   // Request read/write access to TM object,
   // returning current payload
   Object OpenForReadDSTM();
   Object OpenForWriteDSTM();
}
```

Transactional objects are implemented via two levels of indirection (Figure 4.7(a)). First, an application uses an instance of TMObject to uniquely identify each object in the program. Data structures are built by using references to the TMObjects, rather than to the underlying application data. In turn, a TMObject refers to a *locator* for the object, and the locator identifies: (*i*) the transaction that most recently opened the object in write mode, (*ii*) an old value for the object, and (*iii*) a new value for the object. A transaction descriptor itself records a status (ACTIVE, COMMITTED, ABORTED). For simplicity, the locators themselves are immutable, and so updates to these three fields are made by using an atomic CAS on the TMObject itself to replace one locator with another.

An object's *logical state* can be determined using these data structures:

- If the previous writer is COMMITTED, then the locator's new value is the logical state of the object.

- If the previous writer is ACTIVE or ABORTED, then the locator's old value is the logical state of the object.

This definition is the crux of how a transaction can atomically update a set of objects: if its descriptor is linked to all the objects involved, then a single change to its status field from ACTIVE to COMMITTED will change the logical contents of all of the objects. Conversely, if one transaction TA encounters a conflict with another transaction TB, then TA can force TB to abort by changing TB's status field to ABORTED using an atomic CAS operation.

Let us consider the operations invoked by an example transaction, TC, in more detail:

The OpenForWriteDSTM operation (Figure 4.7(b)) starts by checking if TC already has the object open for write—i.e., if the object's locator already refers to TC. If so, then the operation returns the new value from the locator. Otherwise, TC must gain control of the object. First, if the object is currently controlled by another active transaction (say, TD), then TC must ensure that TD has finished. It does this by using an atomic CAS to change TD's status field from ACTIVE to ABORTED. At this point, either TD's status field is ABORTED, or it is COMMITTED (if TD updated its own status, concurrent with TC's CAS). In either case, the outcome of TD is now fixed.

The next step is for TC to take control of the object. To do that, it determines the current logical contents, based on the existing locator, and then allocates a fresh locator with the current object installed as the old copy, and a clone of the object installed as the new copy. TC uses an atomic CAS to replace the old locator with the new one (detecting a conflict if there has been a concurrent replacement of the locator). Note how *none* of these steps changes the logical contents of the object;

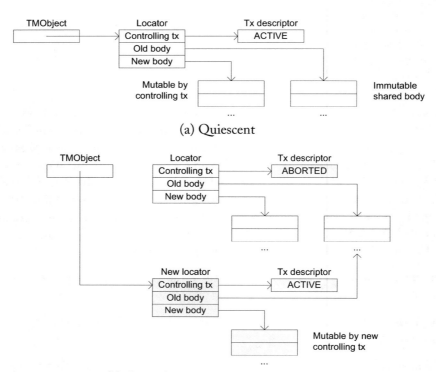

(a) Quiescent

(b) Open for write by a new transaction

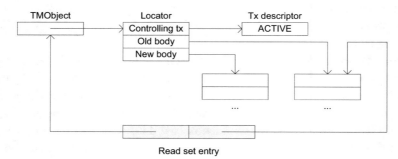

(c) Open for read by a new transaction

Figure 4.7: DSTM data structures and operations.

the new locator will yield the same logical contents as the old one, but places the object under TC's control.

The `OpenForReadDSTM` operation (Figure 4.7(c)) is simpler, determining the logical contents of the object and adding the object to the transaction's read-set (if it is not already held there).

The `CommitTx` operation needs to perform two steps. First, it must validate the read-set, to check that it is still consistent at the point of commit. Second, if validation succeeds, the transaction uses an atomic CAS to change its status field from `ACTIVE` to `COMMITTED`. If the transaction commits successfully, then the start of validation forms the linearization point of `CommitTx`.

DSTM provides opacity, and the original description of DSTM used re-validation of the complete read-set upon each `OpenFor*` operation [146]. Techniques such as a global commit counter [302], visible readers, or semi-visible readers [188; 191] can be used to avoid full validation much of the time.

Although the need to use `TMObject` structures and explicit `OpenFor*` calls can appear cumbersome, Herlihy *et al.* [147] showed how this kind of boilerplate can be generated automatically, leading to a programming model comparable to the `atomic` blocks in Chapter 3. Korland *et al.* used a Java bytecode rewriting framework to add STM instrumentation when loading Java classes, supporting a programming model where individual methods can be marked `@atomic` [174]. These atomic methods are instrumented with STM operations, along with their callees.

4.5.2   NONBLOCKING OBJECT-BASED STM DESIGN SPACE

DSTM was a highly influential early STM system, and it formed the basis for many subsequent designs. The recurring themes in these systems are (*i*) dynamically associating objects with a transaction descriptor, to allow a single update to the descriptor's status field to effect an atomic update, (*ii*) defining a notion of the logical contents of an object in terms of a set of metadata values and object snapshots, and (*iii*) using shared data that is generally immutable, such as locators, so that a thread can determine the object's logical value without needing locking. Examples of this form of STM include those of Scherer and Scott [278], Guerraoui *et al.* [122], Fraser and Harris [105; 106], Marathe *et al.* [209; 211], and Spear *et al.* [302].

This last piece of work introduced a taxonomy of the design choices which these systems take:

- *When are objects acquired for writing?* As in lock-based STM systems, the main options are *eager acquire* in which a writer takes control of an object before writing to it, versus *lazy acquire* in which a writer only takes control as part of its commit operation. Eager acquire enables earlier conflict detection because the presence of a writer is visible. Lazy acquire means that a transaction only causes conflicts when it is ready to commit, which can avoid livelock. DSTM [146] is an example of eager acquire, while Fraser's OSTM [105; 106] is an example of one using lazy acquire. Marathe *et al.* describe an adaptive STM which can select between both choices [209].

- *Are readers visible to writers?* As in lock-based STM systems, invisible, semi-visible, and visible-reader strategies are possible. Invisible readers avoid contention between readers in the memory system. Visible readers enable earlier application of contention-management policies and can avoid the need for incremental validation if opacity is to be provided. Marathe *et al.* describe the use of per-object lists of visible readers [211], and Spear *et al.* describe the use of bitmap-based visible readers, with each possible reader being represented by a bit in a single word added to each object [302].

- *How are objects acquired by transactions?* Generally, either the TMObject refers to another metadata object, such as a DSTM locator, or the TMObject refers to a transaction descriptor from which the current contents are determined. The latter approach is used in OSTM (Figure 4.8), with a TMObject using a tag bit to distinguish whether or not the object is currently acquired. If the object is quiescent then the TMObject refers directly to the current body. If not, then it refers to a transaction descriptor, and the relevant body is found from its read/write-set entries. Avoiding locators reduces the number of levels of indirection from 2 to 1. However, without locators, STM systems usually need to visit each written object twice while committing: once to install a transaction descriptor to acquire the object, and then once more, after commit, to remove the descriptor.

- *How do objects appear when they are quiescent?* The main questions here are how many levels of indirection are used between an object header and its payload, and whether or not fewer levels of indirection can be used when an object is quiescent. Marathe *et al.* describe a technique to let a TMObject refer directly to an object payload when it is not acquired [209]. In subsequent work, Marathe *et al.* inline the new object payload into a locator-like structure. They also introduce a "clean" bit into the TMObject to signal that the transaction descriptor from the locator has committed [211] (Figure 4.9). This bit avoids needing to look at the descriptor (typically reducing the number of cache lines accessed and allowing the descriptor to be re-used).

- *Which nonblocking progress property is provided?* This typically entails a choice between lock-freedom and obstruction-freedom. Lock-free STM systems usually acquire objects lazily, so that if transaction TA finds itself obstructed by transaction TB, then TA is able to help TB finish its commit operation. Lock-freedom can avoid livelock between transactions, but it can introduce heavy contention in the memory system and, although it guarantees system-wide progress, it does not provide guarantees to any given thread.

4.5.3   NONBLOCKING STM SYSTEMS WITHOUT INDIRECTION

The final set of STM algorithms that we examine combine nonblocking progress guarantees alongside aspects of the lock-based designs from Section 4.2–4.4. Like many of the lock-based designs, these STM systems avoid the need to add levels of indirection to the representation of objects in

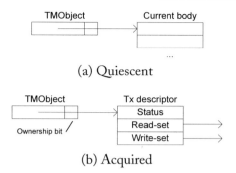

(a) Quiescent

(b) Acquired

Figure 4.8: Memory structures in OSTM.

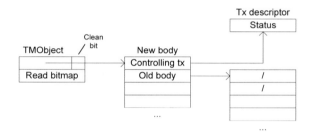

Figure 4.9: Memory structures in RSTM.

Figure 4.10: Memory structures in WSTM.

memory; this can help simplify integration between the STM system and the rest of the language implementation.

Harris and Fraser described the first of these systems [134]. This was a word-based STM system, termed WSTM. With WSTM, data is usually stored in its normal format in the heap, using a separate table of STM metadata entries which are associated with the heap under a hash function

(Figure 4.10). WSTM uses lazy acquire and invisible reading, and each transaction's descriptor holds (`addr,old-val,new-val,version`) tuples for each location that the transaction has accessed. Each metadata entry can hold either a version number, or it can refer to a controlling transaction along with an ownership count (described below). A bit in the metadata distinguishes these cases.

A word W's logical contents is defined in terms of W's value and the value of the metadata word associated with it, $M(W)$:

- If $M(W)$ holds a version number, then W holds its logical contents directly.

- If $M(W)$ is owned by an `ABORTED` or `ACTIVE` transaction, then W's logical value comes from the old value in the owning transaction's descriptor (if it contains an entry for W) and from W directly, otherwise.

- If $M(W)$ is owned by a `COMMITTED` transaction, then W's logical value comes from the new value in the owning transaction's descriptor (if it contains an entry for W) and from W directly, otherwise.

The use of hashing makes this more complex than the object-based systems because a metadata entry acquired for one location will implicitly acquire other locations that hash to the same metadata. The overall design, however, is similar to a lazy-acquire object-based STM in that a commit operation starts by acquiring the metadata for its write-set, then it validates its read-set, then writes its updates, and releases the metadata.

To make this algorithm nonblocking, if transaction TA is obstructed by transaction TB, then TA uses the values in TB's descriptor to help TB finish. There is one difficult case: if TB is in the middle of writing back values to memory, then it will resume these writes when it is next scheduled, potentially trampling on any intervening updates. To avoid this problem, the metadata entries include an *ownership count*, incremented and decremented on each acquire and release. Direct access to a location is enabled only when the count returns to zero; while it is non-zero, any updates must be retained in a transaction descriptor linked from the metadata entry.

Marathe and Moir describe the design and implementation of this form of nonblocking word-based STM in more detail [208], along with many extensions and enhancements to the use of metadata-stealing techniques. They show how to provide variants using eager version management and using global-timestamp-style validation. They show that, although the algorithm is very complicated, the fast paths of a `ReadTx` operation are essentially the same as those of systems like McRT-STM and Bartok-STM.

Harris and Fraser described an alternative implementation using OS support to remove transactions from the commit function [135], and Spear *et al.* showed how Alert-on-Update hardware can be used to avoid this OS support [306] (Section 5.5).

NZTM.   Tabba *et al.* [313; 314] designed a nonblocking object-based STM system which combines ideas from systems like DSTM with zero-indirection access to objects. Figure 4.11 shows the structure of an object in NZTM. Each object can be in one of two modes *inflated*, in which case the

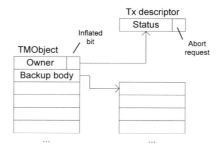

Figure 4.11: Memory structures in NZTM (non-inflated).

object header refers to a DSTM-style locator, or *non-inflated*, in which case the object header refers to a status field for the transaction which controls the object.

While in the non-inflated mode, the operation of NZTM is relatively straightforward. Each object has a *backup copy* which was taken before the currently-owning writing transaction acquired the object. If the owning transaction is ACTIVE or ABORTED then the object's logical contents are taken from this backup copy. Otherwise, if the owner is COMMITTED, the logical contents come from the main object body.

Unlike earlier nonblocking STM systems, aborts are handled co-operatively via an *abort request* flag which is stored in memory along with the owner's status: if a transaction observes its own flag set, then it acknowledges this by updating its own status to ABORTED.

If transaction TA requests transaction TB abort, but it does not receive a prompt acknowledgment, then TA can *inflate* the object rather than waiting indefinitely. Inflation entails setting up a DSTM-style locator and installing a reference to this locator as the owner of the object (along with the inflated bit set). The logical state of the object now comes via the locator: as with the ownership count in WSTM, the underlying state of the object cannot be accessed until the original owner has finished aborting since other threads cannot prevent the original owner from making updates to the original object body.

In practice, inflation would be expected to be rare because it occurs only if the owning transaction is unable to be scheduled for some reason (e.g., a page fault that takes a long time to be serviced).

Tabba *et al.* describe how a single-compare-single-store operation would avoid much of the complexity in NZTM, by letting a thread make all of its writes conditional on its abort-request bit remaining clear [313; 314]. Marathe and Scott describe a similar, but earlier, simplification of WSTM [210].

4.6   ADDITIONAL IMPLEMENTATION TECHNIQUES

The case studies in Sections 4.2–4.4 illustrate many of the techniques used in STM systems for the core problems of data versioning and conflict detection. In this section, we survey the implementation techniques which are needed in order to build a full STM system from these foundations.

First, we describe how to support privatization and publication idioms, focusing in particular on what is needed in order to support race-free programs (Section 4.6.1). Then, in Section 4.6.2, we show how transaction-based condition synchronization can be implemented. Finally, in Section 4.6.3, we describe support for irrevocable transactions.

4.6.1   SUPPORTING PRIVATIZATION SAFETY AND PUBLICATION SAFETY

In order to support privatization idioms, an STM system must respect synchronization that the threads perform: if Thread 1 executes a transaction (TA) that is serialized before a transaction (TB) by Thread 2, then code after TB should also appear to execute after TA. For instance, this might occur if TA accesses an object that is shared, then TB makes the object private, and then code after TB accesses the private object without using transactions. Menon *et al.* call this *privatization safety* [220].

A similar *publication safety* property requires that if Thread 1 executes an operation *before* TA, then the effects of this operation must be seen by TB. However, Menon *et al.* show that publication safety is provided by STM systems so long as (*i*) the language need only support race-free programs, (*ii*) the compiler does not hoist memory operations onto new program paths (such transformations may introduce races into otherwise race-free programs), and (*iii*) the STM system itself does not introduce speculative reads inside a transaction (e.g., via granularity problems such as those in Section 2.2.3).

In this survey, we consider race-free programs, and, consequently, we focus on the techniques for ensuring privatization safety. Work such as Grossman *et al.*'s [115], Menon *et al.*'s [219; 220], and Spear *et al.*'s [299] has shown that there are many alternative and subtle definitions for programming models that support various kinds of benign races.

In supporting privatization safety, the crux of the problem is ensuring that parts of the *implementation* of transactions do not appear to occur after the transaction's position in the serial order—for instance, ensuring that zombie transactions are not visible to non-transactional code, and ensuring that if TA is ordered before TB, then TA must have finished writing back its memory updates before any code after TB accesses those locations.

4.6.1.1   Native Support

Some STM systems provide native support for privatization safety. This occurs naturally in systems that use a commit token to serialize writing transactions—e.g., some configurations of JudoSTM, NOrec, and RingSTM.

In these systems, the serialization of writing transactions ensures that the order in which writers depart from their commit function corresponds with their logical commit order. In addition, the use of global metadata ensures that a read-only transaction, for whom data has been privatized by a previous writer transaction, must wait for the writer to completely commit.

Dice and Shavit's TLRW system [86] demonstrates an alternative approach to providing native support for privatization safety. TLRW is based on encounter-time acquisition of read-write locks and uses pessimistic concurrency control for all data accesses. The locks used by TLRW are designed to be effective on chip-multi-processors in which the lock can be held in a cache that is shared by the threads running transactions.

4.6.1.2 Process-Wide Fences

We now turn to techniques for supporting privatization safety in STM systems which do not provide it natively. The first set of these techniques is based on *process-wide* fences to constrain the order in which transactions finish their CommitTx operations.

As part of the McRT-Malloc allocator, Hudson *et al.* [162] used a *quiescence* mechanism to ensure that a block of memory could be de-allocated: a free operation is deferred until transactions which started prior to the free have finished or been successfully validated. This ensures that there will be no residual accesses to the block, once it has been deallocated, and so it can be re-used for a new purpose. Quiescence was implemented using an epoch-based scheme, with a global variable identifying the current epoch, and transactions publishing the epoch number when they started. An object can be deallocated once the epoch at which the free request was made precedes the current epochs of every current thread. Handling deallocation is effectively a special case of the privatization idiom. Similar quiescence mechanisms have been used in implementations of nonblocking data structures [105].

Wang *et al.* [327] further developed this quiescence mechanism to use a linked list of active transactions. A committing transaction is only permitted to finish once all of the transactions that started prior to the commit have quiesced (either finished their own commit operations or have validated successfully against the committing transaction). This ensures that, once a transaction TA has finished its own commit operation, then (*i*) any transactions which are earlier in the serial order have also finished their commit operations (hence, code after TA will see the effects of these earlier transactions), and (*ii*) any transactions which conflict with TA will have noticed that they are zombies (and finished rolling back their updates, when the technique is employed with eager version management).

Concurrently with Wang *et al.*, Spear *et al.* [301] generalized quiescence to support privatization idioms in general. They introduced the idea of privatizing transactions executing a *transactional fence* which causes them to wait until all concurrent transactions have finished, and hence until the now-private data will not be accessed transactionally.

Menon *et al.* show how privatization safety can be enforced in a TM with lazy version management by a *commit linearization* mechanism [219]. In this technique, transactions obtain unique tickets from a global sequence number at the start of their CommitTx operations. They wait, at the end of CommitTx, for any transactions with earlier ticket numbers to have finished:

```
bool CommitTx(TMDesc tx) {
  mynum = getTxnNextLinearizationNumber();
  commitStampTable[tx.threadID] = mynum;

  if(ValidateTx(tx)) {
    // Commit: publish values to shared memory
    ...
    commitStampTable[tx.threadID] = MAXVAL;
    // Release locks
    ...
    // Wait for earlier TX
    for (int i = 0; i < numThreads; i++) {
      while (commitStampTable[i] < mynum) {
        yield;
      }
    }
  } else {
    commitStampTable[tx.threadID] = MAXVAL;
   // abort : discard & release
   ...
  }
}
```

In this example, `CommitTx` obtains a unique linearization ticket number (`mynum`) and publishes this in `commitStampTable` during its validation and write-back work. Once it has released its locks, it waits until `mynum` is less than any entry in the `commitStampTable`. Consequently, this ensures that transactions finish executing `CommitTx` in the same order that they are linearized, and so any non-transactional accesses after one transaction will see any transactional accesses from earlier transactions.

Marathe *et al.* describe a similar technique for privatization safety with lazy version management used by Detlefs *et al.* [212]. This is a ticket-based scheme using one shared counter that is used to issue tickets to serialize writers, and a second shared counter which indicates which writers are allowed to exit their `CommitTx` operation. A writer acquires a sequence number just before validating its read-set. After its writebacks, it waits for its ticket to be "served", before incrementing the current ticket for its successor. This avoids iterating over a `commitStampTable`.

An attractive property of commit linearization and ticket mechanisms is that they require a transaction to wait only for other transactions to finish executing the `CommitTx` function, rather than requiring waiting while application code executes. However, unlike the earlier quiescence schemes, commit linearization is not sufficient to support privatization safety if eager version management is used: it does not ensure that conflicting transactions have observed the commit and finished rolling back.

On some systems, a disadvantage of process-wide fences is that they introduce communication between transactions which are accessing disjoint sets of data. Yoo *et al.* show that providing privatization safety based on quiescence can be costly and scale poorly as the numbers of threads grows [341].

4.6.1.3 Reader Timestamps

The underlying reason for needing process-wide fences is that, if an STM system uses invisible reads, then a writing transaction cannot determine which other transactions it might need to synchronize with. Conservatively, it therefore synchronizes with *all* concurrent transactions.

Semi-visible reads provide a mechanism to remove much of this synchronization, while still avoiding the need for shared-memory updates for every object in a transaction's read-set. Semi-visible reads allow a transaction to signal the presence of readers, without indicating their identity. Compared with fully visible reads, in which the identity of all readers is published in shared memory, semi-visible reads can be more scalable because a set of concurrent readers need only signal their presence *once* in total, rather than once for each member of the set.

Based on this idea, Marathe *et al.* describe a technique to provide privatization safety for an STM system using a global-clock [212]. The key idea is to extend STM metadata to include a *reader timestamp* (RTS) in addition to the timestamp of the last writer. On a read, the STM system ensures that RTS is no earlier than the reader's start-timestamp (RV), incrementing RTS if necessary. At commit-time, a writer detects the possible presence of conflicting readers by looking at the maximum RTS of the locations that it has written to. Read-write conflicts are only possible if this maximum RTS is later than the writer's own RV. If a conflict is possible, then the writer synchronizes with all transactions that started before its commit timestamp.

Assuming transactions typically access disjoint sets of data, the use of read timestamps reduces the number of times that waiting is necessary. Marathe *et al.* describe important details of the algorithm (such as avoiding self-aborts if a transaction performs a read-modify-write sequence) and optimizations (such as a *grace period* which allows the read timestamps to be managed more conservatively, trading off unnecessary synchronization against the frequency of timestamp updates).

4.6.1.4 Commit Indicators

The final privatization mechanism which we examine are *commit indicators* (CIs) from SkySTM [188]. Unlike read timestamps, CIs can be used with per-object version numbers, as well as with a global clock.

CIs exploit the observation that, if a transaction TW makes an object private, then it must conflict with any other transaction TR that concurrently accesses the object (e.g., via a common location that they access to determine whether or not the object is shared—not necessarily a direct pointer to the object). In a system with lazy version management, Lev *et al.* identify that TW's `CommitTx` operation should not be allowed to complete until all of TR's effects are written back to memory.

Conceptually, CIs extend each metadata entry with a count of currently-committing transactions which have read from data protected by the entry. A transaction increments the CIs for each

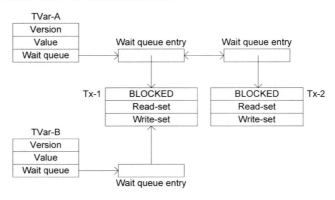

Figure 4.12: Condition synchronization in STM-Haskell.

metadata entry in its read-set as part of its validation. It decrements the CIs after its write-backs are finished. In addition, before releasing write ownership, a transaction waits until the CIs of all of the locations to which it has written have decreased to zero. Lev *et al.* discuss constraints on the ordering between these steps [188] and optimizations to use scalable nonzero indicators (SNZI) [97], rather than simple counters.

4.6.2   CONDITION SYNCHRONIZATION

How do we combine condition synchronization with the STM systems from Sections 4.2–4.5? Semantically, the `retry` operation (Section 3.2.1) can be seen as aborting the current transaction and re-executing it, but a simplistic implementation built this way clearly performs very poorly because of the resources consumed while spinning. Even if hardware threads are plentiful, the STM-level and memory-level contention would usually be unacceptable.

To avoid actually spinning, an implementation needs a mechanism to allow a thread running a transaction to block and to be woken when a possible update occurs. The implementation must ensure that lost wake-ups do not occur (i.e., in the case of `retry`, that a thread is not left asleep once a conflicting update has been committed).

STM-Haskell implements `retry` by installing a form of "trip wire" linking each piece of STM metadata to a list of descriptors for transactions that have accessed it (Figure 4.12). A thread installs these trip-wires before sleeping, a subsequent update notices them, and wakes the threads that they refer to. In STM-Haskell, each transactional variable (TVar) contains a head of a doubly-linked list of wait-queue entries for sleeping threads' transactions. This allows multiple threads to wait on the same TVar and, of course, for a single thread to wait on all of the TVars in its read-set.

In STM-Haskell, the `retry` implementation associates a lock with each transaction descriptor and another with each TVar. A descriptor's lock must be held while changing the state

(BLOCKED/ACTIVE). The TVar's lock must be held while manipulating the list of waiting transactions.

Putting all this together, a `retry` operation:

- Acquires the lock on the current STM descriptor.

- Acquires the locks for the data in its read-set.

- Validates the transaction, to ensure that it (*i*) saw a consistent view of memory and (*ii*) this view of memory is still up-to-date.

- Appends the transaction descriptor to the lists for the read-set locations.

- Updates the transaction's status to BLOCKED.

- Releases all of the locks.

A commit operation briefly acquires the locks for the metadata for each location that it updates, and it wakes any waiting threads linked to them. The STM-Haskell system employs a user-mode scheduling library, and so the actual "sleeping" and "waking" involve directly manipulating the scheduler's data structures. Previous STM systems, without user-mode scheduling, used a per-thread condition variable for sleeping [134].

The STM-Haskell implementation of `orElse` uses the same techniques as `retry`, except that the thread must watch locations in both branches of the `orElse`: any update to any of the locations causes the complete atomic block to be re-executed.

There is an important caveat in the implementation of `orElse` which was incorrect in the original implementation of STM-Haskell. Suppose that the programmer writes this:

```
atomic {
  try {
    // X1
  } orElse {
    // X2
} }
```

If X1 invokes `retry`, then the tentative updates from X1 must be discarded before X2 is attempted. However, the read-set for X1 must be retained and validated when X2 commits—and it must be watched for updates if X2 invokes `retry`. The original implementation simply discarded the read set along with the writes. That is incorrect because the left-to-right semantics in STM-Haskell require that X1's decision to invoke `retry` forms part of the `atomic` action.

4.6.3   IRREVOCABILITY

As discussed in Section 3.2, irrevocability provides a mechanism for integrating TM with some non-transactional features by allowing a designated "unrestricted" [37], "irrevocable" [333] or "inevitable" [308] transaction to run without the possibility of being rolled back. STM systems implement irrevocability by requiring that a transaction acquires a single *irrevocability token* in order to become irrevocable. This can be implemented by an atomic CAS on a shared memory location to indicate which thread, if any, is currently irrevocable.

Welc *et al.* introduced *single-owner read locks* (SORL) to McRT-STM to control the data accesses made by a transaction while it is irrevocable. At most, one thread can hold an SORL on a given object. Holding the SORL prevents concurrent threads from acquiring the pessimistic write lock on the object. However, unlike the write lock, the SORL does not signal a conflict to concurrent readers. This can allow greater concurrency. The extension of versioned locks to support SORL is designed carefully to avoid changing the fast-path operations in McRT-STM [333].

To become irrevocable, a transaction must (*i*) acquire the irrevocability token, (*ii*) acquire the SORLs for all its ordinary read-set entries, and (*iii*) validate that its read-set is still consistent. From that point onwards, the transaction is guaranteed not to experience any conflicts. While executing, an irrevocable transaction must acquire the SORL for each location before it reads from it. In this approach, correctness relies on the fact that all operations which may conflict with an irrevocable transaction will be performed via STM.

Spear *et al.* investigated a range of alternative approaches that trade off the possibility for concurrency alongside an irrevocable transaction, versus the amount of instrumentation needed when irrevocable [308]. In that framework, a global lock-based mode supports irrevocability without any instrumentation of the irrevocable transaction—this enables, for instance, an irrevocable transaction to perform non-transactional uninstrumented operations such as system calls. However, the global lock-based mode does not allow any concurrency between an irrevocable transaction and any normal transactions. Conversely, an SORL-like mode allows complete concurrency between the irrevocable transaction and other transactions, but it requires that irrevocable reads and writes must update shared metadata (either SORLs or a Bloom filter). A third mode allows read-only transactions to commit in parallel with an irrevocable transaction, but it requires non-irrevocable writer transactions to block at their commit point. The various implementations of this last mode include mechanisms that avoid instrumentation of irrevocable reads, as well as mechanisms that require metadata inspection on irrevocable reads but not metadata updates; the mechanisms all require eager acquisition of locations, but sometimes they could do so without an atomic operation.

4.7   DISTRIBUTED STM SYSTEMS

The idea of transactions is often applied to distributed systems, using techniques such as two-phase commit to ensure atomicity between a set of components in the presence of machine failures. It is therefore natural to examine whether the ideas of TM might be applied as part of a distributed programming model and to investigate implementation techniques for distributed STM systems.

4.7.1   STM FOR CLUSTERS

Manassiev *et al.* describe the first distributed STM system [205], targeting clusters of workstations, and building on the underlying Treadmarks software distributed shared-memory system (S-DSM) [72]. The S-DSM system provides low-level page-based synchronization between processes running on distinct machines. In general, contemporary S-DSM systems implement a release-consistency memory model [11], in which updates are propagated at synchronization points. A race-free program sees a sequentially consistent view of memory across all computers.

Manassiev's paper observes that the mechanisms of an S-DSM system can easily be adapted to implement TM. The key differences are that transactions provide a mechanism to specify the execution interval over which memory updates are buffered on a machine before being propagated, and that transactions provide a different semantics for resolving conflicting memory references. A transaction's changes are obtained at a sub-page level by comparing the contents of a transactionally-modified page against the contents of a clean copy. When a transaction commits, the TM sends a change list to other computers which record but do not apply, the changes. Once all computers acknowledge receipt of the change list, the transaction can commit. On a given computer, if the change list updates a page being modified by an active transaction, then the transaction is aborted. If the page is not being modified by a transaction, the changes are applied on demand, the page is subsequently referenced to avoid delaying the committing transaction. Multi-versioning allows a read-only transaction to continue executing with an outdated version of a page, even after receiving an update.

Kotselidis *et al.* investigated a somewhat similar approach at the level of objects within a Java Virtual Machine [179; 180]. They developed a distributed form of DSTM using object-level conflict detection and version management between transactions, and commit-time broadcast of transactions' read/write-sets. When a node receives a broadcast, it compares the read/write-sets with the accesses of its own local transactions, selects whether to abort its local work, or to request that the broadcaster aborts.

4.7.2 STM-BASED MIDDLEWARE

In a series of papers, Cachopo and Rito-Silva [45; 46], Romano *et al.* [267], Carvalho *et al.* [51] and Couceiro *et al.* [71] describe the use of a multi-versioned STM system to manage the in-memory state of a distributed web application middleware. The system is used in the implementation of FénixEDU, a campus activity management system used in several Portuguese universities.

A domain modeling language is used to describe the objects stored in the system and the relationships between them. This model is used to derive code to make the objects persistent. The objects are replicated across a cluster of machines, and transactions are dispatched to replicas under the control of a load balancer. Upon commit, the write set of a transaction is sent to other replicas using an ordered atomic broadcast service. This allows commit/abort decisions to be made locally, given knowledge of the transaction's read/write-sets and of the write-sets of transactions earlier in the serial order.

The use of multi-versioned objects means that read-only transactions can always commit without needing to communicate between replicas. Read-only transactions vastly outnumber updates in the FénixEDU workload.

4.7.3   STM FOR PGAS LANGUAGES

An alternative setting for distributed STM systems are *partitioned global address space* (PGAS) languages, in which portions of an application's heap reside in separate physical address spaces (as opposed to distributed STM systems based on replicating the heap across a set of address spaces).

Bocchino *et al.* designed Cluster-STM for use across systems that might scale to thousands of nodes [41]. In this large-scale setting, eliminating communication is critical for performance, and so Cluster-STM employs techniques to reduce the cases where communication is required, and to aggregate communication when it is necessary. For instance, operations on STM metadata can be piggy-backed on requests to fetch the data itself.

A key design decision is to split a transaction's descriptor across the nodes on which data is accessed, rather than holding the descriptor on the node on which a transaction is initiated. This co-locates the STM metadata with the fragments of all of the descriptors that are accessing it, thereby reducing inter-node communication. Visible reading is used to avoid the need for additional commit-time validation of each data item.

Dash and Demsky extend an object-based distributed STM with a pre-fetching mechanism based on a static analysis to identify "paths" through a series of objects' fields which are likely to be followed by a thread [79].

4.8   STM TESTING AND CORRECTNESS

STM systems are complicated—both in terms of the concurrency-control algorithms that they use and in terms of the low-level implementation techniques which are employed. Errors can creep in at both of these levels, and so proofs of algorithmic correctness and techniques for testing implementations are both valuable.

Lourenço and Cunha [200] illustrated a range of problems that occurred when porting the TL2 STM system to a new platform and the techniques that they used to debug the system. Harmanci *et al.* developed a domain-specific language (TMUNIT) for writing test cases that illustrate problematic interleavings—e.g., examples like those of Section 2.2.3. Manovit *et al.* [206] used pseudo-random testing techniques to exercise a TM implementation and then compared the behavior visible during these runs with a formal specification of the intended semantics.

Guerraoui *et al.* showed that, for a class of STM algorithms with particular symmetry properties, exhaustive testing of 2-thread 2-location transactions is sufficient to identify important properties such as strict serializability, and obstruction freedom [118; 120]. This enables the use of model checking, which is attractive because definitions of an STM algorithm can be kept reasonably close to the source code (e.g., as in the models developed by O'Leary *et al.* for use with the SPIN model checker [235]).

Cohen *et al.* used the TLA+ specification language to model a TM system and the TLC model checker to test finite configurations of it [68]. In later work, Cohen *et al.* incorporated non-transactional accesses and used the PVS theorem prover to obtain a mechanized proof of correctness of a TCC-based system [69].

Tasiran showed that a proof of serializability for a class of TM implementations can be decomposed into proofs that the TM satisfies an "exclusive writes" property, a "valid reads" property, and a "correct undo" property [315]. Based on this, he developed a model of the Bartok-STM system using the Spec# language, and then he used automatic verification to demonstrate that it satisfied these properties.

Doherty *et al.* defined a "weakest reasonable condition" correctness condition for TM and developed techniques to facilitate formal and machine-checked proofs that TM implementations satisfy the condition [89].

Hu and Hutton related the behavior of a high-level semantics for STM-Haskell to a model of parts of the STM implementation on a lower-level abstract machine [161].

Moore and Grossman [226] and Abadi *et al.* [3; 6] provide operational semantics for languages incorporating atomic actions and multiple lower-level models which incorporate aspects of TM implementations—e.g., modeling interleaving between steps of different transactions and forms of conflict detection and version management. For various classes of input programs, these lower-level models are shown to produce equivalent results to the higher-level models.

4.9   SUMMARY

In this chapter, we have introduced techniques for software implementations of transactional memory, grouping systems into those with per-object version numbers (Section 4.2), those with a global clock (Section 4.3), those which use only a small amount of global metadata (Section 4.4), and those which are structured to allow nonblocking transactions (Section 4.5).

At a high level, there are apparent attractions and disadvantages of each approach; it is clear that modern implementations of any of these forms of STM system substantially out-perform early ones, although it is less clear how the relative performance of these mature implementations compares. Inevitably, this will depend greatly on the workload and on the structure of the underlying machine—e.g., whether or not the hardware threads share a common cache, whether or not it is a multi-chip system, exactly what memory model is provided by the hardware and exactly what programming model is supported by the STM system and the programming constructs built over it.

CHAPTER 5

Hardware-Supported Transactional Memory

In this chapter, we introduce the main techniques that have been proposed for hardware implementations of transactional memory. Although STM systems are very flexible, HTM systems can offer several key advantages over them:

- HTM systems can typically execute applications with lower overheads than STM systems. Therefore, they are less reliant than STM systems on compiler optimizations to achieve performance.

- HTM systems can have better power and energy profiles.

- HTM systems can be less invasive in an existing execution environment: e.g., some HTM systems treat *all* memory accesses within a transaction as implicitly transactional, accommodating transaction-safe third-party libraries.

- HTM systems can provide strong isolation without requiring changes to non-transactional memory accesses.

- HTM systems are well suited for systems languages such as C/C++ that operate without dynamic compilation, garbage collection, and so on.

In Section 5.1, we introduce key components of a typical HTM system. We also discuss how these components are implemented using conventional hardware mechanisms. As in the previous chapter's discussion of STM systems, the majority of this chapter is then structured around a series of different kinds of HTM design. The first of these, in Section 5.2, are HTM systems based on conventional mechanisms in which a processor's hardware structures support access tracking and version management, and existing cache coherence protocols detect data conflicts. We conclude the section with a discussion of hardware challenges and the software considerations that they introduce.

Section 5.3 discusses alternative mechanisms that have been introduced to circumvent some of the limitations and challenges of conventional HTM systems—e.g., to reduce implementation complexity by decoupling various HTM functions from a processor's hardware structures. We look, in particular, at systems such as LogTM that use software-managed logs to hold updates [227], systems such as Bulk that use Bloom-filter signatures for conflict detection [53], systems such as TCC that decouple conflict checks from cache coherence protocols [132], and systems such as FlexTM that allow lazy conflict detection via changes to an existing cache coherence protocol [290; 291].

In Section 5.4, we discuss HTMs specifically designed to support *unbounded* transactions—i.e., transactions that may access an arbitrary amount of data or run for an arbitrary amount of time. These systems typically aim to maintain the performance profiles and characteristics of a conventional HTM while removing limits on the transactions that they can run. The proposals for unbounded HTMs vary in the degree to which they combine hardware and software components; a primary motivation for these proposals is to simplify the impact on the software tool-chains and libraries by abstracting TM management from software (similar to how a processor's virtual memory system hides memory management from application software). These goals make such unbounded HTM systems more complex than conventional HTM systems.

A new class of proposals has emerged that suggests exposing hardware mechanisms directly to an STM, rather than executing the transaction directly in hardware. These proposals, discussed in Section 5.5, use the hardware mechanisms to accelerate a SW transaction instead of directly executing code transactionally.

Finally, in Section 5.6, we look at a number of enhancements and extensions which can be applied to many of these different HTM systems. We focus primarily on mechanisms to support nesting in hardware and mechanisms for integrating non-transactional operations (e.g., IO) and system actions.

5.1   BASIC MECHANISMS FOR CONVENTIONAL HTMS

HTMs must perform similar tasks to STMs: they must identify memory locations for transactional accesses, manage the read-sets and write-sets of the transactions, detect and resolve data conflicts, manage architectural register state, and commit or abort transactions. This section discusses the basic mechanisms employed by conventional HTM systems to implement these tasks.

5.1.1   IDENTIFYING TRANSACTIONAL LOCATIONS

The first step for an HTM is to identify transactional memory accesses. This is done via extensions to the instruction set. Based on these extensions, HTMs may broadly be classified into two categories: *explicitly transactional* and *implicitly transactional*:

Explicitly transactional HTMs provide software with new memory instructions to indicate which memory accesses should be made transactionally (e.g., `load_transactional` and `store_transactional`), and they may also provide instructions that start and commit transactions (e.g., `begin_transaction` and `end_transaction`). Other memory locations accessed within the transaction through ordinary memory instructions do not participate in any transactional memory protocol.

Implicitly transactional HTMs on the other hand only require software to specify the boundaries of a transaction, typically demarcated by instructions like `begin_transaction` and `end_transaction`. They do not require software to identify each individual transactional memory access; all memory accesses within the boundaries are transactional.

Explicitly identifying transactional accesses provides programmers with increased flexibility and may aid in reducing the size of a transaction's read-set and write-set. Explicit interfaces also naturally allow transactional and non-transactional accesses to intermix. However, as with most STMs, using an explicit interface can require multiple versions of libraries—one for use inside transactions and one for use outside. Consequently, explicit interfaces can limit the reuse of legacy software libraries inside HW transactions. Therefore, such HTMs are more suitable to construct lock-free data structures where the transactions are fairly small and limited in scope, and software reuse is not a priority. Conventional examples of explicit HTMs include the Herlihy and Moss HTM [148], Oklahoma Update [309], and the Advanced Synchronization Facility [61] proposals.

Implicit identification of transactional locations restricts instruction changes to the boundaries of a transaction. Therefore, these HTM systems can enable reuse of existing software libraries, and they typically do not require multiple code versions to be compiled. However, implicit interfaces provide limited controllability over transactional memory locations. Recent proposals extend implicitly transactional HTMs with a capability to identify *non-transactional* memory locations, thereby bridging the flexibility gap with explicitly transactional HTMs. Examples of implicit HTM systems include Speculative Lock Elision [248], the Rock HTM [82], and the Azul HTM mechanisms [66]. Most modern HTM proposals follow an implicitly transactional model (including most of the proposals discussed in this chapter).

While both HTM models support mixing transactional and non-transactional operations within a transaction, this can introduce specification complexity and memory ordering challenges—especially where the addresses involved may overlap.

5.1.2   TRACKING READ-SETS AND MANAGING WRITE-SETS

As with STM systems, an HTM must track a transaction's read-set and must buffer the transaction's write-set. HTMs have a distinct advantage over STMs in this respect: modern hardware has mechanisms such has caches and buffers already in place to track memory accesses. HTMs can extend these mechanisms to manage their reads and writes without the overheads typically associated with STM systems.

Nearly all modern processors have a hardware cache to exploit spatial and temporal locality of data accesses. Processors look up this cache using a memory address and, if this address is present in the cache, the processor sources data from the cache instead of from main memory. Data access is a performance-critical aspect of processors and so hardware designers optimize caches to provide low-latency access to data. The data cache serves as a natural place to track a transaction's read-set because all memory reads involve cache look-ups.

The early HTM proposal from Herlihy and Moss [148] duplicated the data cache, adding a separate *transactional cache* to track the transaction's read and write sets. Adding a new transactional cache in parallel with an ordinary data cache adds significant complexity to modern microprocessors as it introduces an additional structure from which data may be sourced. It is challenging to add this extra logic to an optimized performance-critical data path. Consequently, extending the existing data cache to track read-sets has therefore become more popular (e.g., this approach is taken in systems such as SLE [248], Rock HTM [82], and Azul's HTM system [66]).

Most proposals based on cache extensions add an additional bit, the *read* bit, to each line in the data cache. A transactional read operation sets the bit, identifying the cache line as being part of the transaction's read-set. This is a fairly straightforward extension to a conventional data cache design. Such designs also support the capability to clear all the read bits in the data cache instantaneously.

Hardware caches enable HTMs to achieve low overhead read-set tracking; however, they also constrain the granularity of conflict detection to that of a cache line. Additional access bits may be added to reduce the granularity for tracking.

Hardware caches can also be extended to track a transaction's write-set. This extension requires the addition of a *speculatively written* state for the addresses involved (e.g., [34; 38; 66; 247]). Since data caches are on the access path for processors, the latest transactional updates are automatically forwarded to subsequent read operations within the transaction; no special searching is required to locate the latest update. However, if the data cache is used to track the write-set, then care is required to ensure that an only copy of a line is not lost—e.g., in systems where caches allow processors to directly write to a location in the cache without updating main memory. A requirement is that the now-dirty cache line must eventually be copied into main memory before it is overwritten by a transaction's tentative updates. If we did not do this, we would lose the only copy of the cache line following an abort.

To avoid any modifications to the data cache, some HTM proposals add dedicated buffers to track read-sets and to buffer write-sets (Oklahoma Update [309] and the Advanced Synchronization Facility [61] are two examples). Hybrid models are also possible; e.g., the SLE proposal and Rock HTM both use the data cache to track the read-set while using the store buffer to track the write-set. Using dedicated buffers may be sufficient if the HW transaction is expected to operate on only a handful of memory locations.

Zilles and Rajwar studied the effectiveness of a data cache for tracking read-sets and write-sets [348]. They considered an HTM design with a 32KB 4-way set associative data cache with 64-byte cache lines and found that such an HTM could frequently support transactions up to 30,000 instructions that operate on hundreds of cache lines; that was fairly large compared to common perceptions. However, in this kind of system, a transaction must be aborted whenever an overflow occurs. Such overflows occur due to set associativity conflicts. Overflows can further be mitigated using a victim cache. A single victim cache increases the data cache utilization for transactions by 16% (increasing utilization to 42%).

A few proposals exist that do not provide a cache line granularity and instead provide hardware support for objects, a software construct. Khan *et al.* [169] integrate transaction management with the object translation buffer used in a processor with an object-based model of execution [334; 335] instead of a conventional cache line model.

5.1.3   DETECTING DATA CONFLICTS

Just as with STM systems, HTMs must detect data conflicts. This typically requires checking whether the read-sets and write-sets for concurrently executing transactions overlap and conflict. Here again, HTM systems have an advantage over STM systems: HTMs that use local buffers to track read-sets and write-sets can use cache coherence mechanisms to detect data conflicts. In this section, we first provide a brief overview of common cache coherence approaches, and then demonstrate how HTMs can build on these coherence protocols to implement conflict detection.

Caching in a multiprocessor system results in multiple copies of a memory location being present in different caches. Hardware cache coherence provides mechanisms to locate all cached copies of a memory location and to keep all such cached copies up-to-date. While numerous cache coherence proposals exist, two common alternatives are *snoop*-based and *directory*-based.

A read operation requires locating the up-to-date cached copy. In a snoop based system, if a processor needs to locate the latest value of a memory address, it broadcasts the request to all other caches. Conversely, a directory-based coherence protocol maintains a directory entry for each memory location, and this entry records where all the copies of a location reside. Many alternatives and hybrids exist, but all require some mechanism to locate all cached copies.

A write operation to a cache must keep all cached copies up-to-date. This usually involves invalidating stale (out-of-date) copies or (less commonly) updating those copies with the newly written value. If two processors simultaneously issue a write to the *same* location with different values, cache coherence protocols ensure that all processors observe the two writes in the same order, with the same value persisting in all copies. In *invalidation-based* cache coherence protocols, the cache with the dirty copy of a line (i.e., the cache that has modified it with respect to memory) is responsible for servicing read requests from other processors which require access to the line.

To support cache coherence, each cache line in a cache has an associated cache *state*. Sweazey and Smith [310] provide a MOESI classification which forms the basis for nearly all coherence protocols. The MOESI classification is based on the various states a cache supports, and the terms arise from the names of the cache states. The *invalid state* (I) means the address is not present in the cache. The *shared state* (S) means the address is present in the data cache and it may also be present in other data caches, hence the shared nature of the state. Because of the shared nature, if a processor needs to update the location, it must request write permissions from other data caches. A processor cannot update an S state cache line by itself. In contrast, for the *exclusive state* (E), an address is exclusively present in the data cache and is not present in any other data cache. A processor can typically update the E state cache line directly without asking for permission. When it does so, the processor changes the state of the cache line to the *modified state* (M). This is the latest copy of the cache line and is only present in this cache. This data cache is responsible for servicing a subsequent request to this line. The E and S states share one common feature: in both cases, the cache line is clean, and the latest copy of the cache line also resides elsewhere, either in memory or another cache. The MESI states are the most common states in commercial multiprocessors. However, some systems also support the *owned state* (O). In the O state, the line is dirty but is also possibly shared with other caches. Since the line is dirty, this data cache owns the line and is responsible for providing this data when other processors request it. However, since it is shared, the processor cannot write the line without requesting write permission from other caches.

While the above mechanisms may seem elaborate and complex, nearly all modern microprocessors support them. HTMs can utilize these existing MOESI protocols to detect conflicts: any write request from another processor to a line in the M, E, S, or O state that has been read transactionally is a data conflict—another processor is attempting to write to a line that has been read in a transaction. Similarly, any read request from another processor to a transactionally-written line in the M state is also a data conflict. In both these cases, the underlying cache coherence protocol automatically provides mechanisms to identify data conflicts.

5.1.4   RESOLVING DATA CONFLICTS

Once a data conflict has been detected, it must be resolved. Nearly all conventional HTM proposals perform eager conflict detection (Section 2.1.3), aborting the transaction on the processor that receives the conflicting request, and transferring control to software following such an abort. This approach is simplest from the point of view of a hardware implementation. A software handler then executes and may decide to reexecute the transaction or to perform some kind of contention management. Notification of the causes of conflicts can be a valuable tool in debugging performance problems in software using TM: most HTM proposals provide a new architecture register for this purpose.

Selecting the ideal hardware contention management technique remains an open research problem, similar to contention management in STM systems. Research has shown the importance of effective conflict resolution—for instance, Shriraman and Dwarkadas's work on the FlexTM system [290] showed that eager conflict detection requires the careful use of contention management in order to remain effective under high contention; we return to FlexTM in Section 5.3.4.

5.1.5   MANAGING ARCHITECTURAL REGISTER STATE

So far, we have discussed tracking and managing state in memory. However, in addition to memory state, software also has associated register state that is architecturally visible.

Collectively, these architectural registers and memory combine to form a processor's *precise state*. From the point of view of software, a single thread in a program has a sequential execution model [296]. A program counter identifies the instruction that the processor is to fetch from memory; the processor starts executing the instruction; the instruction may access memory and may operate on registers in which data may be stored. When the instruction's execution completes, the program counter is updated to identify the next instruction to execute. The register state at this point becomes part of the processor's precise state for the retired instruction. This sequence also occurs within an HTM transaction. When a transaction aborts, the register state therefore must also be restored to a known precise state, typically corresponding to just before the start of the transaction.

Some HTM systems rely on software to perform such a recovery while other HTM systems build on existing hardware mechanisms to recover architectural register state. One straightforward approach for hardware register recovery entails creating a shadow copy of the architectural registers at the start of a transaction. This can be maintained in an architectural register file (ARF). This operation can often be performed in a single cycle. An alternative approach can be possible in processors that already employ speculative execution to achieve parallelism and high performance. These processors rename the architectural registers into an intermediate name to allow multiple consumers and writers of these register names to execute in parallel, while maintaining the correct data dependencies and program order. The intermediate name is often part of a larger pool of registers, also referred to as physical registers to contrast them with architectural registers. A special table, called the register alias table (RAT), tracks the mapping between the architectural registers and physical registers at any given time. For these processors to support register recovery in an HTM, the register mechanisms must be enhanced.

Software managed register recovery can be appropriate if the target transactions are simple and small—e.g., operating on a few data items and not calling into unknown functions. In such scenarios, static analysis can be used to identify registers to save and recover; this analysis becomes more complex with longer transactions, leading to larger save/recover footprints, and consequential overheads. Hardware supported register recovery tends to perform better in such transactions.

Hybrid design are also possible—for instance, relying on hardware to recover the contents of the stack-pointer, frame-pointer, or callee-save registers, and relying on software to recover the contents of the remainder of the register set.

5.1.6 COMMITTING AND ABORTING HTM TRANSACTIONS

Committing a HW transaction requires making all transactional updates visible to other processors instantaneously. In turn, this requires obtaining the appropriate permissions for all locations accessed by the transaction and then ensuring that the local state for these locations transitions to a committed state without being interrupted.

For HTMs that use a local hardware buffer such as a store buffer, this requires obtaining write permissions for all the transactional addresses, blocking any subsequent requests from other processors, and then draining the store buffer into the data cache. Even though the drain is not instantaneous, it is of finite duration, other processors cannot observe any intermediate state.

HTMs that use the data cache to buffer transactional stores require a new cache state array design. Blundell *et al.* describe one approach incorporating flash-clear and conditional-flash-clear facilities (Figure 5.1). This design provides a cache cell where read and write sets can be operated upon instantaneously: when the `clear` signal is asserted, both the read and written bits are pulled down. When `conditional_clear` is asserted, the valid bit is pulled down if the speculatively-written bit is high.

In such designs, the commit and abort operations can also be performed in a cycle. However, cache accesses are on the critical path in modern microprocessors and adding logic into state array

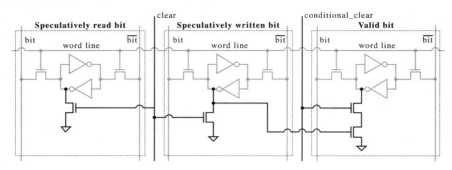

Figure 5.1: Adapted from Blundell *et al.* [38], the figure shows an example SRAM cell augmented with a flash-clear and a conditional-clear capability.

cells introduces timing risk and complexity. Nevertheless, reusing existing structures and data paths is highly desirable.

5.2   CONVENTIONAL HTM PROPOSALS

The first set of HTM designs we look at extend conventional microarchitecture mechanisms to incorporate transactions—they use hardware structures to track accesses, they use buffers or caches to hold tentative state, and they use a conventional cache coherence protocol to detect conflicts. This approach supports transactions of modest size (typically hundreds of cache lines) and modest duration (tens of thousands of instructions). These designs also rely on alternative software mechanisms if a transaction exceeds local cache buffering or its scheduling quanta.

We start by considering conventional HTM proposals where explicit transactional memory accesses are required (Section 5.2.1). We then consider proposals where memory accesses within a transaction are implicitly transactional (Section 5.2.2). Next, we examine hybrid systems that combine an HTM system for some transactions with an STM system for others—e.g., switching to software for long-running transactions, or those that experience frequent conflicts, or those that use features not supported by a given HTM (Section 5.2.3). Finally, with these examples in mind, we discuss the trade-offs in the design and specification of the interfaces to HTM systems (Section 5.2.4).

5.2.1   EXPLICITLY TRANSACTIONAL HTMS

We start by considering four forms of explicitly transactional HTM system—i.e., designs in which new instructions are used to identify which memory accesses should be performed transactionally. Memory locations accessed through normal load and store instructions do not participate in the transactional memory protocol, even if these accesses occur during a transaction.

These HTMs provide programmers fine control over hardware resources of an HTM and often suit programming models geared towards the construction of lock-free data structures.

We start with an influential proposal from Jensen *et al.* [165]. While the proposal provided tracking for only a single address, it may be considered the first optimistic synchronization primitive based on cache coherence mechanisms and laid the foundations for future conventional HTM proposals. The paper observed that a processor could use the cache coherence protocol to optimistically monitor a memory location for conflicts and conditionally perform operations if the location did not experience a conflict. This observation was later generalized by Herlihy and Moss and the Oklahoma Update proposals.

Jensen et al.'s *Optimistic Synchronization.*   The proposal included a reservation register and provided three new instructions to the programmer: `sync_load` for reading a memory location (and initializing the reservation register), `sync_store` for writing a memory location if the reservation was still valid, and `sync_clear` to explicitly terminate the speculation. A coherence conflict to the reservation register would result in a reservation loss and this was communicated through a condition code.

A `sync_load` operation starts monitoring a given location, and a subsequent `sync_store` succeeds only if no conflict has occurred. The store does not need to refer to the same data address as the load, thereby allowing programmers to monitor one location and write another. For example, if the algorithm computes only one data item from a set of data, a programmer convention can establish one data item to define exclusive access over the entire data set.

Jensen *et al.*'s paper influenced synchronization mechanisms in numerous commercial microprocessors; MIPS, Alpha, and PowerPC-based processors implemented variants of the optimistic synchronization proposal [70; 166; 293].

Herlihy and Moss HTM.   Herlihy and Moss coined the term *transactional memory* and introduced three key features for an HTM system: special instructions to identify transactional memory locations, a transactional cache to track the read and write sets of the transaction, and extensions to a conventional cache coherence protocol to detect data conflicts.

Programmers use three new instructions to explicitly identify transactional accesses: `load-transactional`, `store-transactional`, and `load-transactional-exclusive`. The latter provides a way to identify read accesses for which a subsequent write was likely. Programmers can explicitly end transactions using a `commit` or `abort` instruction.

The following code sequence shows the use of Herlihy and Moss's instructions to insert an element into a doubly linked list (taken from [148]). While this usage requires a certain level of programmer expertise, it also provides a flexible interface to identify addresses to track and buffer.

```c
// Usage of new instructions to construct data structure
typedef struct list_elem {
  struct list_elem *next;
  struct list_elem *prev;
  int value;
} entry;

entry *Head, *Tail;

void Enqueue(entry* new) {
  entry *old_tail;
  unsigned backoff = BACKOFF_MIN;
  unsigned wait;
  new->next = new->prev = NULL;
  while (TRUE){
    old_tail = (entry*) LOAD_TRANSACTIONAL_EXCLUSIVE(Tail);
    // ensure transaction still valid
    if (VALIDATE()) {
      STORE_TRANSACTIONAL(&new->prev, old_tail);
      if (old_tail == NULL) {
        STORE_TRANSACTIONAL(&Head, new);
      } else {
        STORE_TRANSACTIONAL(&old_tail->next, new);
        // store pointer transactionally
      }
      STORE_TRANSACTIONAL(&Tail, new);
      if (COMMIT()) // try to commit
      return;
    }
    wait = random() % (01 << backoff);
    while (wait--);
    if (backoff < BACKOFF_MAX)
    backoff++;
  }
}
```

The proposal (Figure 5.2) incorporates a dedicated fully-associative *transactional cache*, in parallel to an existing data cache. The additional cache tracks a transaction's accesses and buffers its tentative updates. Herlihy and Moss proposed exposing the size of the transactional cache to programmers, thereby allowing code to be written with a guarantee that transactions of a given size can execute in hardware.

A TACTIVE hardware flag tracks whether the processor has an active transaction and a TSTATUS hardware flag tracks whether the processor has received an abort condition. The proposal does not include an explicit instruction to start a transaction; instead, a transaction implicitly started on the first transactional instruction executed while the TACTIVE flag is clear.

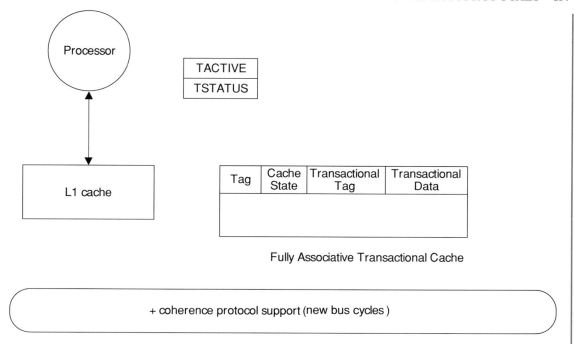

Figure 5.2: Herlihy and Moss's dedicated transactional cache.

Adding a new cache in the data path for loads and stores requires special mechanisms to ensure the correct architectural state is available. When a transactional access is performed, two entries for the address are allocated—one with a *commit tag* (storing the original data value) and one with an *abort tag* (storing the new data value). On a successful commit, the entry with the commit tag would be discarded. On aborts, the entry with the abort tag would be discarded. Maintaining two entries allows the transactional cache to retain an up-to-date copy of a memory location irrespective of whether a transaction commits or aborts. To allow transactional state to spill over from the transactional cache, the authors suggest the use of a "LimitLESS" directory scheme [56], in which software can emulate a large directory. This allows using hardware for the common case and software for the exceptional case.

An ownership-based protocol detects conflicts by matching incoming coherence requests against the transactional cache in parallel with the data cache. If a conflict is detected, then a TSTATUS flag is set to indicate that an abort condition has occurred. A transaction that experiences a data conflict does not abort immediately; instead, it continues to execute, thus admitting inconsistent reads. To ensure that a transaction does not see an inconsistent view of memory (Section 2.2.2), the proposal includes a `validate` instruction which lets a transaction test the status flag. Software must perform validation, recover register state on aborts and ensure forward progress (e.g., using contention management, of the kind described in Section 2.3).

Herlihy and Moss also provide conflict resolution where a processor could selectively refuse giving up ownership (through the use of NACK responses). This improves performance in the presence of conflicts and reduces livelock situations.

Oklahoma Update.   Concurrent with Herlihy and Moss' proposal, Stone *et al.* proposed the multi-word Oklahoma Update [309] mechanism. As with Herlihy and Moss' system, Oklahoma Update was intended for writing scalable shared-memory data structures by allowing short critical sections to be replaced by multi-word atomic updates.

In contrast to Herlihy and Moss' transactional cache, Oklahoma Update extended Jensen *et al.*'s proposal to include multiple reservation registers (up to 8), thereby providing a form of HTM-like multi-word atomic update. It introduced new instructions read-and-reserve, store-contingent, and write-if-reserved to manipulate the reservation registers. These reservation registers also served as storage for the tentatively buffered data and were exposed to the cache coherence protocol to enable conflict detection.

The Oklahoma Update checked for and resolved data conflicts at commit time in two phases. In the first phase, the processor requests write permission for locations in reservation registers that did not have these permissions. Deadlocks may arise during the permissions acquisition phase. To avoid deadlocks, the hardware obtains write permissions in ascending order of address [67]. If the incoming request address is larger than the least reserved address for which the processor does not have write permissions, the processor releases its reservation. If the incoming request address is smaller than the least reserved address for which the processor does not have write permission, the processor defers the request in a buffer and services it later. This prevented livelock but did not provide starvation-freedom or wait-freedom. Once all permissions have been obtained, the second phase starts and the processor commits the data values. During this phase, the processor does not abort and is uninterruptible. The two-phase implementation was similar to two-phase locking from database transaction-processing systems [32]. The processor services the deferred external requests on commits and aborts, servicing the first request to a given address and then forcing a retry of the remaining requestors queued to the same address.

Advanced Synchronization Facility.   Recently, the Advanced Synchronization Facility (ASF) proposal [61] from Advanced Micro Devices takes a similar approach to the explicit HTM systems discussed so far. It introduces a SPECULATE instruction to begin a transaction, along with a COMMIT instruction to mark the end. Control returns implicitly to the SPECULATE instruction if the speculative region aborts, setting the processor flags to indicate that this has occurred. Simple flattened nesting is supported; speculation continues until the outermost COMMIT occurs. ASF proposes the use of a LOCK prefix to be added to memory accesses that should be performed transactionally. In the implementation proposal, ASF proposes the use of dedicated registers, similar to Oklahoma Update, to perform a multi-word compare-and-swap-like operation.

Neither the Herlihy and Moss proposal nor the Oklahoma Update proposal perform register recovery in hardware but rely on software to recover register state. The ASF proposal recovers the stack pointer but relies on software to recover the remaining registers. This software-centric approach for register recovery is motivated by the target usage model. Since these proposals are geared towards fairly small-sized transactions that target lock-free data construction and require explicit identification of transactional locations, they already require a certain level of programmer expertise and code modification. Requiring programmers to also take care of architectural state recovery would seem to be a minor incremental burden.

5.2.2 IMPLICITLY TRANSACTIONAL HTM SYSTEMS

In implicitly transactional HTMs, a programmer uses new instructions to identify the boundaries of a transaction, and all memory accesses performed within those boundaries are considered to be transactional. In this kind of model, the programmer does not have to explicitly identify transactional memory accesses.

Speculative Lock Elision.　Speculative Lock Elision (SLE) [248; 249] is an implicitly transactional HTM design proposed for a modern out-of-order processor. The SLE proposal extends a processor's existing hardware mechanisms to perform access tracking and version management (using the existing data caches and write buffers for doing so). Unlike Herlihy and Moss' HTM, the SLE proposal re-uses the speculative execution engine's mechanisms to roll back register state, rather than relying on software. In addition, SLE identifies transactional accesses based on instruction boundaries instead of individual memory locations; within a transaction, all accesses are transactional.

SLE uses the data cache to track transactional read-set by extending each cache line with an access bit. The SLE implementation uses the store buffer to hold updates performed transactionally. Aborts require the store buffer's tentative updates to be discarded. Commits require making the store buffer's updates visible instantaneously. The hardware issues exclusive ownership requests for the addresses in the store buffer but without making the tentative updates visible. SLE also sets the access bit along with the M state when a tentative store accesses the data cache. This enables conflict checks against stores. SLE cannot initiate the commit sequence until all such requests have been granted. Once all required coherence permissions are obtained, the tentative updates must be made visible atomically. SLE stalls incoming snoop requests from other threads while it is draining tentative updates from its store buffer into the data cache. This ensures all updates are made visible instantaneously. This is guaranteed to complete in bounded time and without global coordination.

Different variants of SLE have examined approaches for recovering register state in an HTM system. One option was to restrict the transactional state to that maintained by the speculative execution engine (e.g., the reorder buffer size [296]). This allowed reusing the existing mechanisms for misspeculation recovery but also limited the size of transactional regions. In the register checkpoint approach, either the architectural register file would need to be saved (through a flash copy mechanism) or enhancements made to the physical register file management mechanism to ensure the physical registers that were architectural at the start of the region are not released and remain available for recovery.

The SLE software model is quite different from the Herlihy and Moss HTM. Instead of providing a new programming interface, SLE re-uses the existing lock-based programming model.

The key observation introduced by SLE was that a processor does not actually have to acquire a lock, but it need only to monitor that it remains available. The processor executes the critical section optimistically as if it is lock-free: two threads contending for the same lock could execute and commit in parallel without communication, so long as their executions were data-independent. It takes advantage of a key property of a typical lock variable: a lock-release's write undoes the changes of the lock-acquire's write. By eliding both of these operations, but monitoring the lock variable for data conflicts, the lock remains free and critical sections can execute concurrently.

The processor starts SLE by predicting that a given atomic read-modify-write operation is part of a lock-acquire. A predictor and confidence estimator determines candidates for lock elision by inspecting the synchronization instructions used to construct locks. The processor issues the read of this read–modify–write operation as an ordinary load, and it records the load's address and the resulting data value. The processor also records the data for the write of the read-modify-write operation, but it does not make this write operation visible to other processors, and it does not request write permission for the cache line holding it. This has two effects: (*i*) it leaves the lock variable in an "unlocked" state and the cache line in shared state, allowing other processors to see the lock as available, and (*ii*) the write value allows detection of the matching lock release by watching for a store by a corresponding unlock operation. This provides program-order semantics: subsequent accesses to the lock variable from the processor performing the elision will return the last written value in program order.

This approach allows SLE to be compatible with existing programming models without requiring new instructions to be used (indeed, the implementation retains binary compatibility with existing software). However, SLE requires the program's implementations of locks to follow a particular pattern that SLE can predict. If optimistic execution cannot continue (e.g., due to lack of cache resources or IO operations), the buffered write data of the atomic read-modify-write is made visible to the coherence protocol without triggering a misspeculation. If the coherence protocol orders this request without any intervening data conflicts to either the lock or the speculatively accessed data, then the execution is committed. Here, the execution transitions from a lock elision mode into an acquired lock mode without triggering a misspeculation.

In this model, a data conflict would cause execution to re-start from the instruction identified to start the critical section, except elision would not be attempted. This way, the exact same program sequence could be executed both with and without SLE.

While the SLE proposal used prediction to identify transactional regions, this required software to follow a specific pattern and limited software flexibility. Using new instructions or software annotations treated as hints may circumvent this limitation [247].

In SLE, conflict resolution was rudimentary, and the transaction receiving a conflicting request would always abort. Transactional Lock Removal [249] extended SLE to use timestamp-based fair conflict resolution to provide transactional semantics and starvation freedom, using Lamport's *logical clock* construction [186] and Rosenkrantz *et al.*'s *wound-wait* algorithm [268]. In the algorithm, a transaction with higher priority never waits for a transaction with lower priority. A conflict forces the lower priority transaction to restart or wait. The algorithm provides starvation freedom, and guarantees that at least one transaction from each set of conflicting transactions does not abort. This hardware support provides significant performance gains in the presence of high contention, even compared with the best software queued locking

Rock HTM.   Rock HTM is an implicitly transactional HTM designed for a modern processor from Sun Microsystems [58]. The Rock HTM's instruction set interface consists of a `chkpt <fail-address>` instruction, where the `fail-address` specifies an alternate target instruction address for the processor to resume execution following an abort [82]; all operations between the `chkpt` instruction and the subsequent `commit` instruction are transactional.

Similar to SLE, Rock HTM also uses the data cache to track the read set of a transaction and used the processor's store buffer to buffer transactional writes. However, Rock HTM requires the level two (L2) cache to track all store addresses inside the transaction. When the processor executes the `commit` instruction, its L2 cache locks all the cache lines corresponding to the tracked store addresses. The stores then drain the store buffer and update the cache. The last store to each line releases the lock on that line. This locking of the cache lines prevents conflicting accesses during the commit sequence.

The Rock HTM implementation exploits a flash copy mechanism to save the architectural register state. Konstadinidis *et al.* provide the implementation details of this mechanism [172]. The register file is a combination of a static memory-cell portion (to reduce area) and an active register-file portion (for speed). Checkpoints are stored in the static portion, which is implemented using a 4-column SRAM array. The active portion maintains the register file used for execution (4-write and 7-read ports for a multithreaded design). The checkpointed register state can be restored into the active portion in a single cycle. Depending on the usage, a write to a register location can independently update the active portion as well as the checkpointed portion. For HTM usage, the checkpointed portion is restored immediately into the active portion. A similar register checkpoint capability has also been employed in an in-order processor [80].

The Rock HTM provides a `cps` register which holds a bitmap that identifies reasons for a transaction failing. These include resource overflows, transient events (such as interrupts), coherence conflicts, and processor exceptions. Microarchitecture limitations in the HTM implementation mean that various events cause transactional aborts—e.g., TLB misses, register-window `save/restore` instructions, a divide instruction, and certain forms of control transfer. Aborts caused by these events were quite problematic in practice, requiring code to be aware of the constraints on its execution. A key lesson from the Rock HTM is to ensure an HTM design does not abort in a manner that is hard to predict and reason about.

Compared with, say, Herlihy and Moss HTM, Rock HTM places fewer demands on the architectural specification because transactional execution is not guaranteed to succeed, and hence a software fall-back path must always exist (entered by branching to the software handler that was registered with the chkpt instruction). For use with lock elision, the fall-back can be as simple as re-executing the same code path directly.

The Rock HTM adopts a slightly different approach than SLE to lock elision. It exploits the same principle as SLE regarding the lock variable and the need to read the location inside the hardware transaction. If the lock is free, speculation can continue. If it is held, the HTM would abort execution and spin. By reading the lock variable, the HTM adds it to the transaction's read-set, and so, if another thread writes to the lock, the HTM would detect the conflict and abort. A key difference between the SLE model and the Rock HTM model is how the lock variable itself is treated. Under SLE, if the lock variable is explicitly read in the transaction, it returns a locked value (the same as in execution without SLE). In the Rock HTM model, if the lock variable is read in the transaction, the read returns a free value (a value different from what it would be if lock elision had not been attempted). The following example shows the usage:

```
txn_begin handler-addr;
if (lock) { txn_abort;}

// body of critical section
Y = X + 5;

txn_end;
```

Note the contrast with the SLE model. Here, two new instructions are added to replace the LOCK/UNLOCK pair, the lock is explicitly tested as part of the critical section, instead of implicitly as part of the LOCK instruction, and a software handler must be provided in case speculation fails. The body of the critical section remains the same as with a lock-based critical section.

Azul HTM. The Azul HTM system [66] is also an implicitly transactional HTM that uses the data cache to track accesses and buffer transactional state. However, it relies on software to perform register recovery.

Large Transactional Memory (LTM). While the conventional HTMs so far abort when a transactional line is evicted from the cache or the store buffer is full, Ananian *et al.* [15] proposed an extension where evicted transactional lines in an HTM would be moved by hardware into a dedicated area in local memory without aborting. Their HTM system, called Large Transactional Memory (LTM, Figure 5.3) allocates a special uncached region in local memory to buffer transaction state that spills from the cache. This region is maintained as a hash table. Each cache set has an associated overflow bit (O bit). This bit is set when a transactional cache line (tracked using a T bit) is evicted and moved into the hash table. During this process, the processor responds to incoming snoops with a negative acknowledgment (NACK). The processor does not service incoming snoops until it has checked both the cache and the memory hash table for a conflict.

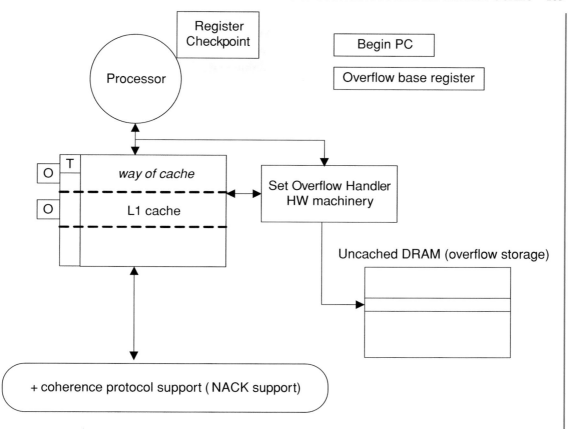

Figure 5.3: LTM design to overflow cache lines into DRAM.

When a request from another processor accesses a cache set with its overflow bit set but does not find a match for its request's tag, a hardware state machine walks the hash table to locate the line with a matching tag. If a processor's request hits in the cache and the O bit is not set, the cache treats the reference as an ordinary cache hit. If the processor request misses in the cache and the O bit is not set, the cache treats this request as an ordinary cache miss. If the O bit is set, even if the request does not match a tag in the cache, the line may be in the hash table in memory. The processor has to walk the hash table in memory. If the walk finds the tag in the hash table, the walker swaps the entry in the cache set with the entry in the hash table. Otherwise, the walker treats this as a conventional miss. If a processor executing a transaction receives a snoop request that conflicts with a transactional line (the line's T bit is set), then the transaction is aborted. In LTM, any incoming snoops that hit a set with the O bit set but do not match the tag interrupt the processor and trigger a walk of the hash table. All these actions occur completely in hardware with no software knowledge or involvement. The area in DRAM can be viewed as an extension of the data cache. In LTM, two independent programs that do not share any data and are in their own virtual address spaces may interfere with each other. Because a transaction in one program can continually interrupt another transaction in another program by virtue of simply running on the same system, the system does not offer performance isolation [347].

Ananian *et al.* describe a register recovery implementation based on a physical register file [15]. In this approach, a new `xbegin` instruction identifies the start of a transaction and causes a snapshot to be taken of the register alias table (RAT). The RAT maintains mapping information between architectural and physical registers. Then, when the `xbegin` instruction retires, the RAT snapshot is retained. To ensure the physical registers named in this snapshot are not released, a new S-bit is added to each physical register, indicating whether or not the register is part of an HTM snapshot. If the S bit is set, the physical register is added to a register-reserved list instead of a conventional free-list. When the instruction committing the transaction retires, the active snapshot's S bits are cleared and the register-reserved list drains into the register-free list. On aborts, the RAT snapshot is used to recover the register state.

Access tracking and conflict detection techniques have been implemented within a cache directory by Titos *et al.* [316]. Ordinarily, a directory records which processors have cached copies of a given line. To detect transactional conflicts, the directory must also be aware of whether or not each processor is executing transactionally, and, if so, which lines are part of the read-set and write-set. In Titos *et al.*'s design, this requires a processor to either flush its cache on starting a transaction, or for transactions to report their cache hits to the directory.

5.2.3 HYBRID TMS: INTEGRATING HTMS AND STMS

In hybrid TM models, the HTM is used to provide good performance, while an STM serves as a backup to handle situations where the HTM could not execute the transaction successfully—e.g., for transactions which require features that are not supported by a given HTM implementation (e.g., pre-emption, or IO). This approach may reduce the pressure on HTM implementations to provide such features. In hybrids, however, care is needed to ensure that the resulting system's semantics are acceptable: for instance, strong isolation is lost if an HTM that provides it is combined with an STM that does not.

These hybrid models require the HW and SW transactions to be able to co-exist, often through arranging that they monitor common memory locations (similar to the way in which locks are monitored when using HTM for lock elision). Example techniques include monitoring the status of a given transaction (e.g., a transaction descriptor's status word), whether or not any SW transaction is executing (e.g., an overflow count), or details of the particular memory locations being managed by software (e.g., a set of ownership records). There are trade offs involving the number of additional locations that must be accessed by HW transactions, the possibility of false conflicts between HW and SW transactions, and the number of separate copies of code that must be compiled (e.g., whether a single copy can be retained, whether separate HW and SW copies are needed).

In the discussions below, the Rock HTM implementation is considered as a canonical reference point for HTM systems.

Lie's Hybrid TM. Lie proposed one of the first hybrid TM systems [194], using it as a way to support unbounded transactions with a simple HTM. In Lie's design, the HTM transaction's code is extended to include checks to detect conflicts with SW transactions. The STM itself is an object-based system implemented in Java, somewhat similar to Herlihy *et al*.'s DSTM system (Section 4.5). In Lie's design, if an STM transaction writes an object, it replaces the original value with a special FLAG value and updates the object in a separate buffer. Hardware transactions use this value to detect conflicts.

When a non-transactional read encounters FLAG, it aborts the transaction that was modifying the object, restores the field's value from the most recently committed transaction's record, and re-reads the location (taking into account the possibility that the location actually should hold FLAG). A non-transactional write aborts all transactions reading and writing the object and directly updates the field in the object. If the program actually intends to write FLAG, then the write is treated as a short transaction to ensure proper handling.

HyTM based on the Rock HTM. Damron *et al*. [78] proposed a subsequent hybrid TM system, similar to Lie's but using a line-based STM as a fall-back. In Damron *et al*.'s design, transactions execute either using an HTM, or as SW transactions using an ownership-table-based STM. Since the HTM uses processor buffers for access tracking and version management, detecting conflicts due to SW transactions is handled through the conventional coherence protocol. However, special care is required to ensure the HTM does not violate isolation of SW transactions. Specifically, HW transactions need to check the STM's ownership-table metadata for any addresses that they access; the HW transaction aborts if the ownership table indicates the presence of a conflicting software transaction.

A code example is shown below. This is a transformation from the HTM example above (from [78]):

```
tx_begin handler-addr
if (!canHardwareRead(&X)) { txn_abort; }
tmp = X
if (!canHardwareWrite(&Y)) { txn_abort; }
Y = tmp + 5
txn_end
```

The canHardwareRead and canHardwareWrite are functions provided by the HyTM library:

```
bool canHardwareRead(a) {
  return (OREC_TABLE[h(a)].o_mode != WRITE);
}

bool canHardwareWrite(a) {
  return (OREC_TABLE[h(a)].o_mode != UNOWNED);
}
```

As we can see, there is no lock associated with the code sequence. Therefore, the HTM code path requires checks against an STM data structure to ensure correct interaction. While this approach requires only a basic HTM for the STM to co-exist, it adds additional lookups into the HTM code path thereby inflating the potential read-sets and adding overhead. Baugh *et al.* study the impact of this and show that this behavior can, at times, seriously affect performance.

Baugh *et al.* studied contention management in the context of Hybrid HTM systems and made some key findings. They found that it was important to have a good contention management policy in hardware. Performance was always better if conflicts never caused a fail-over to software compared to failing-over following some fixed number of aborts or by avoiding HW transactions to abort except when absolutely necessary. This would typically require some form of contention management support; the paper implements an age-ordered scheme where a younger transaction would abort and re-execute. This result also confirms the observation of the TLR paper [250]. The paper also found the benefit of avoiding falling over to STM simply for conflicts since STM transactions with their overhead would extend the duration thereby increasing contention. Schemes that immediately re-execute conflicting HW transactions as SW transactions can cause a cascading effect resulting in all HW transactions into software thus hurting performance even more (an observation also made of the Rock HTM system [82]).

Phased TM.   Lev *et al.* [193] proposed a further form of hybrid design known as *phased transactional memory* (PhTM). Instead of requiring hardware transactions to check for conflicts with concurrent software transactions, PhTM prevents HW and SW transactions from executing concurrently. The PhTM system maintains a counter of currently-executing SW transactions. Every HW transaction checks this counter, and if non-zero, the HW transaction aborts itself. Since the counter is also read inside the HW transaction, any subsequent modifications to this counter also trigger an abort. This approach reduces overheads for HW transactions, but it results in increased aborts (as discussed by Baugh *et al.* [28]): an overflow of even a single HW transaction aborts *all* other concurrently executing HW transactions. To avoid remaining in SW mode, PhTM deploys another counter that maintains the number of running transactions that failed due only to HW limitations. When this counter is 0, PhTM shifts back to a HW phase by stalling the transaction on conflicts instead of re-starting them in an SW mode.

Kumar's Hybrid TM.   Kumar *et al.* [183] proposed a hybrid TM system similar to Lie's that executes a transaction in HW mode and then switches to an object-based STM. However, the hardware design proposed is different from the earlier ones. Two new structures are added. A transactional buffer records two versions for a line: the transactionally-updated value and the original value. A hardware transaction state table has two bits per hardware thread. These bits record if the thread is executing a transaction and if the transaction started in a software or hardware mode.

Each line in the cache has two bit-vectors recording reads and writes, one for each hardware thread. The HTM uses these vectors to detect conflicts: conflicts between two hardware transactions are resolved by aborting the transaction that receives the conflicting request. A conflict between a hardware transaction and a non-transactional request aborts the transaction.

When a hardware transaction detects a conflict and aborts, the hardware invalidates all transactionally written lines in the transactional buffer and clears all read and write vectors. The abort sets an exception flag in the transaction state table but does not abort execution right away. The abort is triggered and a software abort handler invoked when the transaction performs another memory operation or tries to commit.

For the HTM system, Kumar *et al.* use an interface similar to conventional implicit designs – begin/end/abort, and a special instruction to register the abort handler – but they extend this to include non-transactional loads and stores. To support the SW transactions, they add additional instructions (start a SW transaction, explicit load and stores similar to the explicit HTM systems). The paper also extends the DSTM objects and uses the HTM hardware to monitor an object's locator's state. This allows other transactions to abort the SW transaction.

NOrec Hybrid TM.   Dalessandro *et al.* describe how the NOrec STM system (Section 4.4.2) can be integrated with Rock-style HTM [77] to produce a design in which HW transactions need only monitor a single location (in addition to their own data accesses).

NOrec's existing global versioned lock is retained and used for synchronization between SW transactions; in the hybrid design this is known as the *main* lock. In addition, a second versioned lock is used to serialize SW transactions with HW transactions; this is known as the *software* lock.

A SW transaction executes as normal except that, at commit time, it uses a 2-word HW transaction to acquire the software lock alongside the main lock (leaving odd values in both). After writeback, it uses normal writes to release the software lock and then the main lock (restoring both to even values).

A HW transaction starts by reading the main lock and spinning until it is available. Immediately before commit, a HW transaction reads and increments the main lock by 2. This update to the main lock causes any concurrent SW transaction to be re-validated (but, due to NOrec's use of value-based validation, the update does not necessarily cause the SW transaction to be aborted).

The fact that a HW transaction reads from the main lock ensures that a HW transaction does not start concurrently with write-backs from a SW transaction (the atomicity of the HW transaction forces the HW transaction to abort if a SW transaction subsequently commits). By separating the software lock from the main lock, the hybrid NOrec design reduces the window of opportunity for conflicts to occur because of two HW transactions accesses to the software lock. The use of value-based validation enables short-running HW transactions to commit without forcing non-conflicting long-running SW transactions to be aborted.

The hybrid NOrec design illustrates how a limited-size HTM can be useful in implementing parts of an STM system, in addition to running transactions that are written by an application programmer.

5.2.4   SOFTWARE AND DESIGN CONSIDERATIONS

HTM Software Abstraction.   So far we have discussed HTMs that rely on conventional hardware mechanisms. While reusing hardware existing mechanisms is attractive, it also creates an abstraction challenge. Hardware resources are finite, and their implementation is typically abstracted away from the software through the instruction set architecture (ISA). The ISA shields programmers from implementation artifacts and provides a consistent behavior across processor families, serving as a contract between the hardware and software—e.g., it shields the programmer from details such as cache sizes and geometry, cache coherence mechanisms, speculative execution, multiprogramming, context switch intervals and thread scheduling. This approach allows a hardware implementation of a fixed ISA to vary—from relatively simple microprocessors to highly complex high performance microprocessors.

To understand this more clearly, consider a load instruction. From an application program's perspective, this is an ordinary load to a given virtual memory address. However, behind this simple instruction interface, a series of complex operations are being performed by the hardware and system software. The hardware calculates the proper address (depending on the addressing mode of the instruction), checks to ensure the application has sufficient permissions to access and operate on the address, and determines the actual translation of the virtual memory address to where it sits in physical memory. In the absence of a valid translation, it invokes system software to create a translation. Once the hardware has a translated address, it then performs the operation to find where in the memory system the location currently resides, and it reads the location, while ensuring the value returned is coherent and consistent. Of course, modern processors are designed such that these operations are low latency and a load operation typically completes in a few cycles in the common cases. However, none of these intermediate steps is actually visible to the application programmer. This is the abstraction the ISA provides; one can imagine the software complexity and overhead if the application programmer was responsible for performing all these operations.

While this kind of abstraction is fundamental, it introduces interesting challenges for HTM systems, especially since HTM systems try to implement a programmer-visible TM software model using mechanisms that typically are part of the implementation. This often requires a software component to a conventional HTM system to form an additional abstraction layer. This software layer sits above the ISA, similar to an operating system or a language runtime system. However, such an approach limits the benefits of the HW mechanisms to only when the transaction works within the resource limitations imposed by the hardware (e.g., the cache size limiting access tracking) or underlying execution environment (e.g., the scheduling quantum beyond which one thread is de-scheduled and another thread uses the same hardware resources). Numerous extensions discussed in the next sections attempt to improve the quality of this software abstraction.

Progress Guarantees.   The HTMs discussed so far fall into two categories with regards to forward progress: ones that guarantee a minimum resource size available for a transaction to make forward progress (Herlihy and Moss, Oklahoma Update, and ASF) and others that do not provide any guarantee (Rock HTM, and SLE).

However, where guarantees are available, they typically relate to transactions that run in isolation. This raises an interesting question of what such guarantees really mean. If the guarantee is merely that of sufficient hardware buffers, software must still provide mechanisms to ensure forward progress. For example, an external interrupt may always occur in the middle of an HTM transaction, or another thread may execute in a pathological pattern that always triggers aborts by writing to a cache line that is read by a given HW transaction.

In contrast, in current processors, a synchronization instruction such as a compare and swap is guaranteed to complete, regardless of the level of contention or rate of interrupts. Processors implement these instructions in a manner to make them non interruptible. They do so by simply locking down the memory address in a local buffer, operating on it, and then releasing it. This is possible because operating on and locking a single memory location does not create a deadlock, there is no circular dependence possible with other processors. However, HTMs provide the capability to operate on multiple memory locations; ensuring atomicity on multiple locations requires hardware to lock down multiple locations, and resolving deadlocks in a multiprocessor system would require a complex protocol. It is unclear whether a hardware system can provide the same forward progress guarantees of existing synchronization primitives, such as a locked compare and swap, to an instruction set that supports atomic updates on multiple locations. It is an open question whether software can use instructions with weak forms of guarantees without providing an alternate path that does not rely on these instructions.

Diestelhorst *et al.* discuss the challenges of implementing the AMD ASF proposal on an out-of-order x86 core simulator [87], and the approaches they took to guaranteeing eventual forward progress for transactions that run without contention and that access at-most 4 memory locations.

Hardware Complexity Trade-offs. Extensions to support HTM can add further complexity to critical, high-complexity aspects of modern processors; as we discuss in the next section, this has led to alternative implementations that rely less on modifications to conventional processor structures.

Given the numerous open research questions with the TM programming model and the challenges it poses to existing software tool-chains and libraries, it is yet unclear what the most suited hardware support may be. To reduce this risk, Hill *et al.* suggest deconstructing TM mechanisms into distinct and interchangeable components instead of a monolithic implementation [152]. In addition to aiding hardware refinement, this approach allows usages beyond the TM model, such as reliability, security, performance, and correctness. While such a deconstruction is attractive, it can introduce architecture specification and validation challenges. Hardware vendors try to minimize the state space of an instruction specification to provide a well defined contract to software and to make validation tractable. Deconstruction requires individual components to maintain the contract, irrespective of software usage and interactions with other software and hardware features. For example, if conflict detection is made a self contained component, the rules of its use must be well defined, including situations where multiple pieces of software may be expecting to use the same hardware, without being aware of each other.

The HTMs discussed in this chapter vary greatly in their hardware requirements and their complexity. This is most often a direct function of the intended software model. Models that aim to minimize impact on tools, support legacy software libraries, and reduce the extent of software analysis and compiler optimizations often require greater hardware support than models where software re-writing and re-compilation are acceptable, compiler optimizations are effective, and legacy software is less of a concern. It is an open research question where hardware support eventually lands in this wide spectrum.

5.3   ALTERNATIVE MECHANISMS FOR HTMS

The HTM systems discussed so far extend a processor's hardware structures to track accesses and perform version management. This, however, introduces complexity into an already complex design. Further, using processor structures places a limit on the transaction sizes.

This section discusses some of the alternative implementations proposed. Section 5.3.1 focuses on HTM designs such as LogTM [227] where version management is through the use of a software-resident log instead of hardware structures. Section 5.3.2 discusses signatures in systems such as Bulk [53] that compactly represent read-sets and write-sets in a transaction instead of using the data cache. Section 5.3.3 and Section 5.3.4 discusses HTMs that do not use conventional cache coherence protocols for conflict checks, e.g., TCC [132] and FlexTM [290; 291], which can both support lazy conflict detection.

5.3.1   SOFTWARE-RESIDENT LOGS FOR VERSION MANAGEMENT

Moore *et al.* [227] introduced the use of software-resident logs for version management in an HTM. Their proposal of LogTM used a software-resident log instead of hardware buffers. During a transaction's execution, hardware uses this log to record the older values of the memory location being over-written inside the transaction. On a commit, this log is discarded and on an abort, a software handler executes, and restores the original values into memory by walking the log. In this model, hardware does not have any notion of speculative writes; all writes update memory, whether they occur inside or outside a transaction. It is the responsibility of the LogTM coherence protocol enhancements to ensure other threads or transactions do not observe any speculative writes.

The LogTM implementation optimizes commit operations: by updating memory directly during transactions, it did not need to perform any copy operations during commits. This is similar to STM systems that use eager version management (Section 4.2).

Figure 5.4 shows the LogTM hardware. LogTM still uses the data cache to track transactional accesses. It extends each cache line with read (R) and write (W) bits. The R-bit tracks whether this line has been read within the transaction and the W-bit tracks whether this line has been written in the transaction. Since access tracking is linked to the data cache, LogTM does not support transactions that experience a context switch. Each thread has its own software log allocated in virtual memory. LogTM provides a system software interface to initialize a thread's transactional

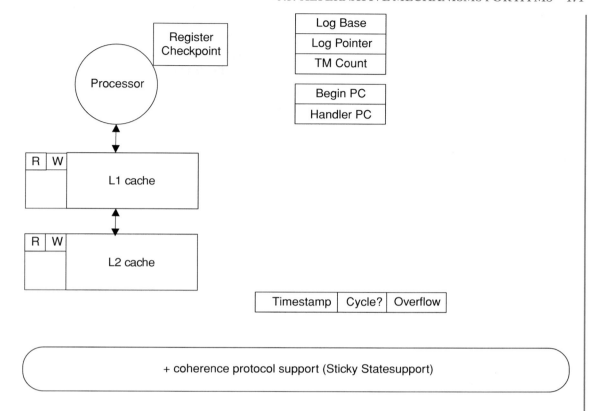

Figure 5.4: LogTM hardware organization.

state, log space, and per-thread conflict handlers, and to perform rollback and aborts. The LogTM system communicates the bounds of the per-thread software-managed log to hardware.

In LogTM, every write in the transaction may add to the software log. To ensure this occurs with high performance, LogTM provides hardware support for a processor to add entries directly into the executing thread's log without invoking any software. When a store operation occurs inside a transaction, the LogTM hardware appends the previous value of the cache line being overwritten by the store and its virtual address to the thread's log. The W-bit is set to avoid repeated logging operations. A single-entry micro-TLB is used to pre-translate the log's virtual address to a physical address that can be used directly by the processor.

When a transaction aborts, control transfers to a software handler (running non-preemptibly). This handler then walks the log in reverse and restores memory state. When a transaction commits, the pointer into the log is reset so that the transaction's entries are discarded. LogTM maintains the software log even if the transaction fits in the local cache.

Moore *et al.* describe various optimizations to reduce logging overhead [227]. The hardware could buffer the updates to the log and not actually write to memory unless it is necessary; the writes to the log can be delayed until either an abort or an overflow. This would mean that if a transaction commits without overflowing the cache, then LogTM would never write to the actual log. Lupon *et al.* suggest a similar optimization where the store buffer performs version management (similar to SLE and Rock HTM) and if the transaction exceeds the store buffer size, then the HTM uses a LogTM-like software-resident log for version management [204].

Recording values in a software-managed log is attractive since it decouples versioning from hardware, and it enables inspection of speculative state by software. While concerns remain about the performance overheads of pure STM systems, the LogTM architecture obviates the need for software handlers to execute during successful transactions by enabling hardware to operate directly on the log. This has the effect of making the log a part of the processor architecture's definition.

By decoupling version management from the data cache, LogTM opened up additional opportunities for access tracking. One approach (by Moore *et al.* [227]) discussed next exposes transactional access tracking into the coherence protocol. Another approach by Blundell *et al.* extends access tracking through the use of an additional tracking structure called the permissions cache [34].

Sticky Coherence State.   In LogTM, a transactional line can escape the data cache without aborting the transaction. This is achieved through the addition of a "sticky" coherence state. This sticky state extends transactional access tracking into the coherence protocol. Even though a line may be evicted, the evicting cache continues as the cache line's owner. This way, transactions use existing cache coherence mechanisms to detect conflicts when a transaction overflows its cache. When a transactional cache line is evicted from the cache, the directory state is not updated and the cache continues to be associated with the line; it becomes `sticky`. This is unlike a conventional protocol where a writeback of a modified line results in the directory assuming ownership of the line.

Consider eviction scenarios for the M, S, O, and E cache-line states. Eviction of a transactionally-written cache-line (M state) results in a transactional writeback instead of a conventional writeback: (*i*) the directory entry state does not change, (*ii*) the processor remains the owner of the line, and (*iii*) an overflow bit is set at the processor. When the directory receives a request for this line, it forwards the request to the owner. If the owner receives a request for a line that is not present in the cache, and the owner has its overflow bit set, then the owner signals a potential conflict by sending a negative acknowledgment (`NACK`) to the requester. The requester then invokes a conflict handler; it does not interrupt the processor that responded to the request.

A protocol that supports silent evictions of clean-shared cache lines (S state) works without special actions: the evicting processor will receive invalidation requests from the directory for the cache line. If the protocol does not support silent evictions, then a sequence similar to that for the M state ensures correct handling by preventing the directory from removing the processor from the sharing vector.

A cache line in the owned O state means that the data value is not the same as in the main memory, but the line is potentially shared. Here, the cache writes the line back to the directory and transitions it to an S state. LogTM treats a silent E eviction as if the line was in M state. The processor must interpret incoming requests unambiguously; it can no longer simply ignore requests that do not match in the cache. This is because commit operations do not actively clean directory state; these states are cleaned lazily. Suppose that processor P1 receives a request to an address not in its cache. If P1 is not in a transaction and receives a request for a line in a sticky state, then the line must be from an earlier transaction. If P1 is in a transaction, but its overflow count is zero, then the request must also be from an earlier transaction. In both cases, P1 responds to the directory with a CLEAN message.

The requester performs conflict resolution after receiving responses from other processors. The requester sends a request to the directory. The directory responds and possibly forwards it to one or more processors. Each processor then inspects its local cache state and responds with either an ACK (no conflict) or a NACK (a conflict). The requester collects this information and then resolves the conflict. Instead of the requester immediately aborting on receiving a NACK, it may reissue the request if the conflicting processor completes its transaction. However, to ensure forward progress and avoid unnecessary aborts, LogTM uses a distributed timestamp method, and it invokes a software conflict handler if a possibility of a deadlock arises. Such a deadlock may arise because a transaction may be simultaneously waiting for an older transaction and may force an older transaction to wait.

Since the access tracking is maintained by hardware (using the data cache in conjunction with the cache coherence protocol), LogTM does not support transactions that experience pre-emption.

Permissions Cache.   Blundell *et al.* [34] provide an alternative approach to extending access tracking in HTM systems that use software-resident logs for version management. They augment the HTM with a *permissions-only cache* to hold information about evicted data. When a cache block is evicted with its *R*-bit or *W*-bit set, then the address and read/write information is allocated in the permissions-only cache. The permissions-only cache tracks the transaction's access information but not the data values. The new cache is accessed as part of external coherence requests.

Commits and aborts clear the new cache. The paper proposes an efficient implementation using sector-cache techniques. Alternatively, the second level cache can be extended to support permissions-only lines. Since the proposal allows transactionally-written lines to escape from the data cache into the next levels of the memory system, it requires a mechanism to restore the transactionally-written data when an abort occurs. Conflict checking with such blocks still occurs via the permissions-only cache.

The paper also proposes an HTM extension to support overflows from the permissions-only cache without aborting; however, only one such transaction is allowed at any time, and no other thread is permitted to execute concurrently (either in a transaction or out of one). A process-wide transaction-status word tracks whether or not an overflowing transaction is active. Since no other thread is executing, this overflowing transaction cannot abort due to data conflicts.

5.3.2 SIGNATURES FOR ACCESS TRACKING

The HTM designs discussed so far still rely on a processor's hardware structures for access tracking, whether it is the data cache, an extra transactional or permissions cache, or special states in the coherence protocols. An alternative approach is to use *signatures* to summarize a transaction's data accesses. Ceze *et al.*'s Bulk system [53] introduced the use of read and write signatures to track the read and write sets of a transaction, specifically to avoid changing the cache design to track transactional accesses.

In these signature-based approaches, each transaction maintains a *read signature* for its read-set and a *write signature* for its write-set. The signatures themselves are fixed-size representations, much smaller than a complete list of the reads and writes of a modestly-sized transaction. While this reduces the state to track read/write sets, it also causes the signature to be conservative (i.e., a signature might suggest that a transaction has accessed a location X, when in fact it has not).

A signature is implemented as a fixed size register. When a processor performs a transactional read, the address is added to the read signature, and on a transactional write, the address is added to the write signature. All HTM proposals based on signatures use some form of Bloom filters (similar techniques have been explored in STM systems, as discussed in Section 4.4.1). However, many variants have been investigated, and the design trade-offs for a hardware implementation are different to those for a software implementation.

Sanchez *et al.* [276] discuss various designs, shown in Figure 5.5. They classify a "true" Bloom signature as one based on a single Bloom filter. Such a true Bloom filter consists of an m-bit field accessed with k independent hash functions (Figure 5.5(a)). The initial value is 0. To insert an address, k hash functions are applied to the address to generate the k bits in the m-bit signature to be set to 1. Similarly, to check whether an address intersects with the signature, the k hash functions are applied and if any of the bits the functions point to is 0, then the address does not intersect. If all bits the functions point to are 1, then either the address is a match or is a false positive.

Sanchez *et al.* point out that implementing k hash function signatures using an SRAM with k read and write ports is area inefficient: the size of the SRAM cells needed increases quadratically with the number of ports. They then describe a parallel Bloom filter implementation (Figure 5.5(b)). This can perform similarly to a true Bloom filter but without the area inefficiency. Instead of k hash functions, parallel Bloom filters have k bloom filters each with m/k-bit fields, each with a different hash function. Inserting an address requires hashing it to set a bit in each of the k filters. This design uses smaller single-ported SRAMs instead of larger multi-ported SRAMS.

Sanchez *et al.* also propose a new "cuckoo" Bloom signature. This combines elements of cuckoo-hashing, which provides an exact representation of the signature's contents when the number of items in the signature is low, along with the use of Bloom filters to provide unbounded storage.

Bulk HTM. Ceze *et al.*'s design employed parallel Bloom signatures with bit-selection hashing. The hardware supports additional operations to assist with conflict detection—e.g., to test whether two signatures have any addresses in common, the hardware implements an intersection function. Similarly, a signature-union combines the addresses from multiple signatures into a single signature.

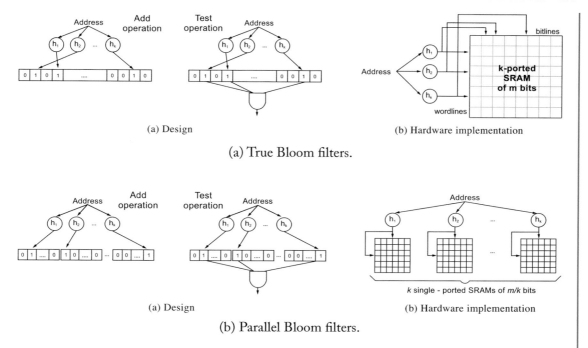

(a) True Bloom filters.

(b) Parallel Bloom filters.

Figure 5.5: Alternative forms of Bloom filter (adapted from Sanchez *et al.* [276]).

This is used when address sets of nested transactions have to be combined. Hardware supports testing an address for membership in a signature. This is used to allow the signature to coexist with a conventional cache coherence protocol and to provide strong isolation if non-transactional read and write operations from other processors conflict with the transaction. Figure 5.6 shows the typical operations on signatures for the Bulk HTM proposal.

Since Bulk continues to use the data cache for version management (even though this is transparent to the cache itself), significant complexity is introduced during commits and aborts to handle the versioned data: the processor must either invalidate and discard transactional state, or make this state visible atomically. To do so, it must identify which lines in the cache are transactional.

Bulk does not record access information in the cache lines themselves, and so this information must be derived from the read/write-signatures. Determining the addresses from a signature is a two-step process. First, the hardware extracts the cache sets corresponding to all the addresses in the signature. Then, a membership function is applied to each valid address in the selected cache sets. A decode operation on a signature extracts an *exact* bitmask corresponding with the sets in the

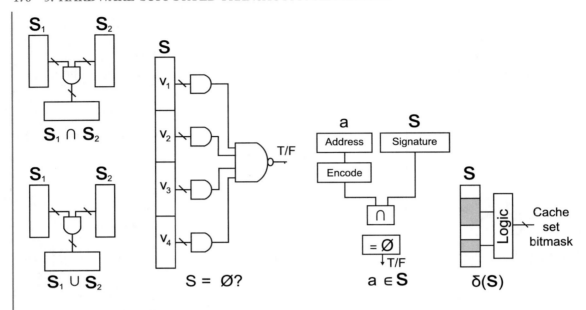

Figure 5.6: Bulk signature operations (adapted from Ceze *et al.* [53]).

cache. For each selected set in the bitmask, Bulk reads the addresses of the valid lines in the set and applies the membership operation to each address.

To commit a transaction, Bulk HTM uses a global commit protocol and signature broadcasts. Before a processor can commit, it arbitrates system-wide for commit permission. Once it obtains permission, it broadcasts its write signature to other processors (unlike the address-based broadcasts described in Section 5.3.3). When a processor receives the write signature, it performs an intersection operation between its local read signature and the incoming write signature. This operation detects whether or not a conflict could have occurred. If the receiving processor aborts, it uses its local write signature to identify and invalidate speculatively-modified lines in its cache. To avoid extending each cache line with a bit to track speculatively written lines, Bulk requires that if a set has a speculatively modified cache line, then that set cannot have any nonspeculatively modified cache line. This restriction prevents an invalidation of a nonspeculatively modified cache line in the set— this could happen because a signature is an inexact representation of addresses and may alias to include nonspeculatively modified cache lines. The signature, however, expands to an exact representation of cache sets with speculatively modified cache lines. This signature expansion is shown in Figure 5.7.

If the receiving processor does not abort, any lines that were written to by the committing processor must be invalidated. The write signature of the committing processor is used to identify these lines. If a cache line receives an ordinary coherence invalidation, then the hardware performs

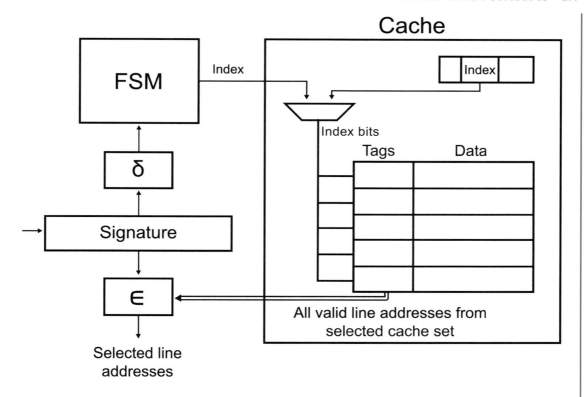

Figure 5.7: Bulk signature expansion (adapted from Ceze *et al.* [53]).

a membership operation of the incoming address on the local signature to check for a data conflict. Execution aborts if a match occurs.

Bulk supports conflict detection at word granularity without changing the granularity of the cache coherence protocol. If word granularity is used for conflict detection, then Bulk must merge updates made to different parts of a cache line on different processors—e.g., one processor that is committing and another processor that receives the write signature. For this, Bulk uses special hardware to identify a conservative bitmask of the words the receiving processor updated in the cache line. It then reads the latest committed version of the cache line and merges the local updates with the committed version. This way, it retains its updates while incorporating the updates of the committing processor. Bulk does not require cache-line bit extensions for this.

In Bulk HTM, transactions are still limited to the size of the data cache and an alternate mechanism is required to support larger transactions.

LogTM-SE.   Yen *et al.*'s LogTM-SE [337] also uses signatures for access tracking and conflict detection. Figure 5.8 shows the typical signature operations in LogTM-SE. LogTM-SE supports

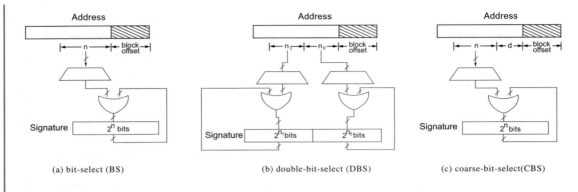

(a) bit-select (BS)   (b) double-bit-select (DBS)   (c) coarse-bit-select(CBS)

Figure 5.8: LogTM-SE signatures (adapted from Yen *et al.* [337]).

operations on signatures similar to Bulk: adding an address to a signature, checking whether an address aliases with a signature, and clearing a signature's "O" set. Unlike Bulk, LogTM-SE continues to use the existing cache coherence protocol for conflict detection, rather than relying on separate global broadcasts of signatures or tokens. Conflicts are detected by comparing the address in an incoming coherence request against the local signatures. We discuss LogTM-SE in more detail later in Section 5.4.1.

5.3.2.1 System Implications of Signatures

While signatures are attractive because they compress information tracking, they introduce numerous subtle interactions and side effects.

Signatures can lead to performance pathologies that are not present in other TM systems. First, signatures introduce a risk of false positives on conflict detection: there is a tension between the size of the signature's representation, the way it is constructed, and the occurrence of false positives. In order to reduce the likelihood of these false positives, a number of researchers have investigated mechanisms to avoid including accesses to thread-local data in signatures [277; 338]. Doing this can reduce the number of addresses that are inserted into the signature and, consequently, reduce the number of false positive conflicts that are produced.

The second performance problem with signatures is that since they are usually based on physical addresses, false conflicts can occur between two processes' signatures, even if their actual address spaces are physically disjoint. This can lead to the risk of denial-of-service attacks in which a malicious process could attempt to engineer conflicts with the transactions occurring in other, unrelated processes. The addition of address-space identifiers can prevent this kind of interaction— but it may also prevent the legitimate use of cross-process transactions.

The use of physical addresses can also require care if a page is relocated: any locations that the signature holds for the old page must be identified, and the corresponding addresses must be added for the new page mapping. (Of course, an HTM system may simply not support such re-mappings within a running transaction).

5.3.3 CONFLICT DETECTION VIA UPDATE BROADCASTS

The HTMs discussed so far use a conventional cache coherence protocol to detect access conflicts. This often means the conflict point is determined at the time the request is detected by the coherence protocol. Alternate proposals exist that decouple conflict checks from conventional cache coherence protocols. In this section, we look at mechanisms that rely on global broadcasts of conflict information. Then, in Section 5.3.4, we look at hardware mechanisms for deferring conflicts.

In update-broadcast systems, if a transaction reaches its commit point successfully, then it notifies other processors of its write set. Conversely, if a processor receives a write-set notification which conflicts with its own read-set, then the processor aborts its own transaction. This means that conflicts are *only* detected against transactions that are committing. A key difference between these HTMs and the ones discussed so far is these HTMs allow multiple concurrently executing transactions to write the same address concurrently. This does not cause conflicts since conflict checks (and write-backs) are deferred until commit time.

Knight's System. Knight [171] proposed a hardware system with elements of this approach as far back as 1986. The hardware system (Figure 5.9) focuses on speculative parallelization of a single-threaded program into multiple threads and then executing the program on a multiprocessor. This is different from typical HTM systems that focus on explicitly parallel threads, but similar to several speculative parallelization proposals that leverage TM (e.g., [324]). In Knight's design, a compiler divides the program heuristically into a series of code blocks called transactions. The hardware executes these transactions and enforces correct execution by detecting any memory dependence violations between the threads. Knight's proposed hardware used two caches: one to track potential memory dependencies (access tracking), another to record the read-set and to buffer speculative updates (version management).

When a processor is ready to commit its transaction, it would wait to receive a "commit token", indicating that all of the older transactions (in program order) have now committed. Upon receiving the commit token, the processor broadcasts its cache updates (writes) to all other processors. Other processors match the incoming writes to their own reads and writes (tracked in the dependency cache). On a conflict, the processor aborts the transaction and restarts.

TCC. Hammond *et al.*'s [132] Transactional Coherence and Consistency (TCC) is similar to Knight's. As with Knight's proposal, TCC requires the program to be divided in a series of transactions—either a single threaded program is divided into a series of transactions, or a programmer may specify explicit transactions in a multithreaded environment.

Unlike traditional HTM systems, the TCC system does not perform conflict checking with an ownership-based coherence protocol; instead, the accesses and changes are locally buffered with no coherence requests. When a transaction is ready to commit, the processor arbitrates for a global

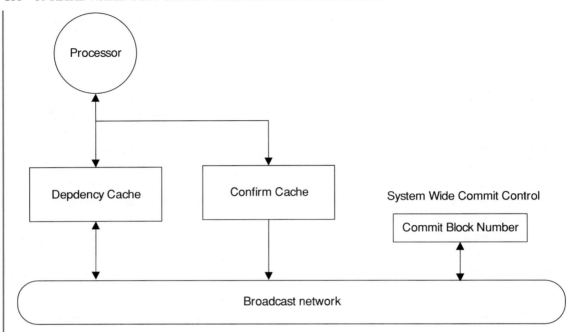

Figure 5.9: Knight's hardware organization.

commit token. Once the processor obtains the token, its transaction commits and the transaction's writes are broadcast to all other processors. All other processors compare the incoming writes to their own transaction's read sets. If a match occurs, the other processor aborts and reexecutes its transaction. If the transaction exceeds local cache buffers, then TCC provides a non-speculative mode. In this mode, the processor executing the transaction retains the global commit token to prevent any other processor in the system from committing; this lets the transaction executes nonspeculatively directly updating memory.

The TCC model unifies two known techniques: speculative parallelization of sequential programs (*ordered transactions*) and optimistic synchronization in parallel programs (*unordered transactions*). While the concept is similar to Knight's, the hardware systems are different (Figure 5.10). TCC proposed numerous approaches to recover register state (suggesting the use of shadow checkpointing, rename-table checkpointing, and a software scheme). TCC also uses hardware caches extended with additional bits. TCC used write buffers to track speculative updates instead of the caches. The R-bits in the cache maintain the read-set, and multiple bits per line mitigate the problem of false sharing. The M-bits in the cache track the write set. The Rn-bits, one for each word/byte in a cache line, are set if a store writes all parts of the word/byte. Subsequent loads to such words/bytes do not need to set the R-bit because the processor previously generated the read data and therefore

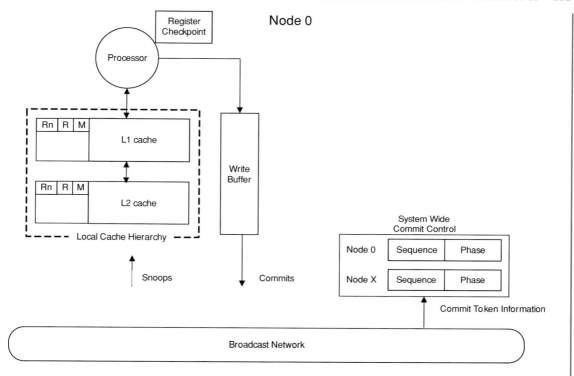

Node 0

Figure 5.10: TCC hardware organization.

cannot depend on another processor. TCC employs double buffering to prevent unnecessary stalls while a processor waits for a prior transaction to commit. Double buffering provides extra write buffers and extra set of read and modified bits for the caches.

Ceze *et al.*'s Bulk HTM (discussed earlier) also employs a similar broadcast approach, where a transaction arbitrates for a global commit token prior to broadcasting its write signature.

The global broadcast serves as a bottleneck for large processor count systems. Chafi *et al.* enhance the TCC system to create what they call Scalable-TCC [55], which avoids the global broadcast but retains the commit-time conflict checks in a directory based memory system. A commit operation requests a unique transaction identifier (TID) from a global co-ordinator. The TID identifies the transaction's position in a serial order. Each directory holds a "now serving" identifier (NSTID) for the most recent transaction which has committed from that directory's point of view. For the directories in a transaction's read set, the transaction sends a "probe" message to check that the directory has served all of the earlier TIDs. For the directories in a transaction's write set, the transaction sends a "mark" message identifying the addresses involved, waiting until the transaction's TID is the directory's NSTID. The transaction also multi-casts a "skip" message to directories in neither set, so that these directories may advance their own NSTID values. Note that TIDs are assigned at commit time, and so it is unnecessary for one processor to wait while another processor executes user code.

Pugsley *et al.* investigate an alternative form of parallel commit in a TCC-like system [245]. Their design avoids the need for central TID allocation, and the need for all transactions to send "skip" messages to directories whose data they have not accessed. Instead, they provide a form of short-term locking of directories, allowing a transaction to "occupy" all of the directories it has accessed. Order-based acquisition of directories, or timestamp mechanisms, can be used to avoid the risk of deadlock in this operation. Once all of the necessary directories are occupied, invalidations are sent for any readers of the locations being written by the transaction. The "occupied" bits can then be cleared. In effect, this is a form of two-phase locking at a directory level.

Waliullah and Stenström [325] designed extensions to a TCC-like system to avoid the risk of starvation when one processor continually commits transactions which conflict with longer transactions being attempted by another processor. The basic idea is to maintain a "squash count" of the number of times that each processor has lost a conflict while attempting its current transaction and to defer an attempt to commit if the processor's squash count is less than that of any other processor.

Broadcast approaches as discussed in this section are new mechanisms that either replace or co-exist with cache coherence protocols. Cache coherence and memory consistency is one of the most complex of hardware aspects to verify and validate, especially in high performance processors that sustain high memory bandwidth and nonblocking caches.

5.3.4 DEFERRING CONFLICT DETECTION

In STM systems, researchers have explored how different conflict detection techniques can provide robustness against several of the performance pathologies of Section 2.3. Intuitively, requiring that one transaction aborts only because another transaction has committed successfully provides a useful system-wide progress guarantee, even though it does not ensure that any individual transaction makes progress. The challenge in applying these ideas in an HTM based on a conventional cache-coherence protocol is that the underlying mechanisms are built around an assumption that each line has at most one writer at any given time. As in STM systems with eager version management, this assumption precludes multiple writers running speculatively.

The FlexTM and EazyHTM systems explore techniques to permit multiple speculative writers to the same location and to perform commit-time conflict detection (similar to the update-broadcast systems discussed above):

FlexTM [290; 291] combines a number of separate features which, collectively, support a mainly-hardware TM system. We return to several of its components in later sections. To support commit-time conflict detection it provides a "programmable data isolation" mechanism (PDI, extended from earlier work [289]), and hardware-managed "conflict summary tables" (CSTs).

PDI is responsible for isolating the tentative work being done by a transaction. It provides TLoad/TStore operations for performing transactional loads and stores, and the implementation

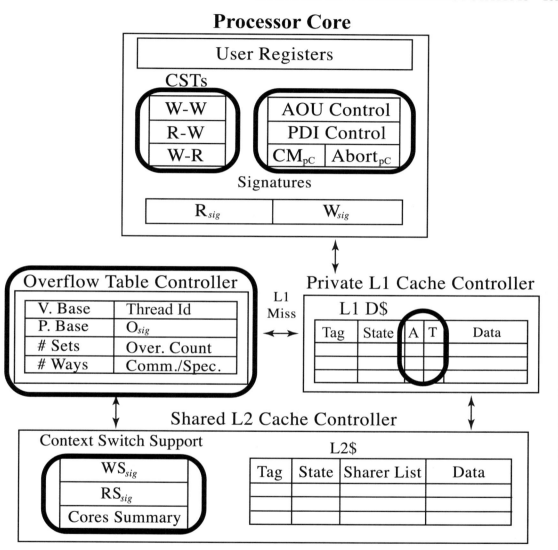

Figure 5.11: FlexTM hardware organization (adapted from Shriraman *et al.* [291]).

extends a cache coherence protocol to allow multiple processors to have tentative transactional stores to the same location at the same time, rather than allowing at most one writer. This is explored in more depth later in Section 5.5.2.

CSTs are responsible for tracking the conflicts that each processor experiences. Each processor has three CSTs: one to track write-write conflicts, one to track conflicts between local reads and remote writes, and one to track conflicts between remote reads and local writes. Each CST holds an entry for each of the other processors in the system. Entries in the CSTs are set eagerly, while a transaction runs, and so the presence of a bit indicates a possible conflict, rather than a definite conflict with a committed transaction.

To support eager conflict detection, software can register a conflict handler which is invoked as soon as a CST bit is set. Consequently, if an eager transaction reaches the end of its execution, then all of the CST entries must be clear and it can proceed to commit.

To support lazy conflict detection, a transaction examines its CSTs at commit time in a software commit function. If a transaction running on processor P1 finds a CST entry for a transaction running on processor P2 then P1 attempts to abort P2's transaction by using an atomic-CAS on a status field for P2's transaction. If P1's transaction successfully aborts any conflicting transaction, then P1 can commit so long as its own status field does not indicate that it has been aborted.

In principle, this design admits the risk of two transactions aborting one another if they attempt to commit conflicting updates concurrently. However, as in STM systems [300], this might be expected to be unlikely, and more elaborate commit routines could implement a contention management policy.

The EazyHTM system exploits similar observations in a directory-based pure-HTM system [318]. Extensions to the coherence protocol are used to inform each of the processors that is sharing a cache line of the presence of the others. Each processor maintains a "racers list" of other processors with which its current transaction has experienced a conflict. The list at one processor, P1, over-approximates the set of transactions which need to be aborted when P1's transaction commits— e.g., a conflict with processor P2 may have been resolved because P2's transaction may have aborted because of a separate conflict with a third processor. To filter out unnecessary aborts, each processor maintains a "killers list" of which processors are currently permitted to abort its transaction. The killers list is populated when a possible conflict is detected. It is cleared upon commit or abort, and a processor ignores abort requests from processors which are not present in its killers list.

Approaches that decouple conflict checking from existing cache coherence protocols do so at a significantly increased hardware complexity. While an attractive concept, providing software direct control of cache coherence mechanisms dramatically increases hardware system validation efforts.

5.4   UNBOUNDED HTMS

So far, we have discussed HTM designs that focus on support for transactions that typically fit in hardware buffers and that do not exceed their scheduling quanta. However, when such limits are exceeded, these HTMs fall back to an alternative execution mode that either results in the loss of transactional properties (e.g., acquiring locks, serializing execution) or involves invoking an STM. Using an STM as a fall-back introduces the ecosystem challenges associated with STMs.

Another class of HTMs, called Unbounded HTMs, propose HTM designs that can support transactions that exceed hardware resources, whether they be storage or scheduling quanta. Their primary goal is to abstract the TM implementation from software and make it non-invasive and high performance, the same way virtual memory is abstracted from application software, and provide a consistent and well defined behavior. These goals makes such HTM designs more complex than the ones discussed earlier and more inflexible than their software counterparts.

Unbounded HTM proposals vary greatly in their scope and implementation complexity. We start with LogTM-SE that combines signatures and software-resident logs to completely decouple transactional state management from hardware. We then discuss a series of proposals that use persistent metadata to achieve the de-coupling. This architecturally defined meta-data is either per memory block or maintained in an alternative software space. Finally, we discuss proposals that extend the page table mechanisms to perform access tracking.

5.4.1   COMBINING SIGNATURES AND SOFTWARE-RESIDENT LOGS

Yen *et al.*'s LogTM-SE [337] combines signatures to track accesses and detect conflicts, with a LogTM-style software-managed log to manage versions (Figure 5.12). This approach avoids some of the complexities associated with using signatures in Bulk: commits only require clearing the signature and, as in LogTM, resetting the log pointer.

In contrast to the LogTM per-cache-line W-bit for filtering logging, LogTM-SE cannot avoid writing a log entry if the address is already set in the signature. This is because signatures permit false positives, and so it is unclear whether or not the particular address has already been logged. Instead, LogTM-SE uses an array of recently-logged blocks as a log filter; this is simple and effective in practice. Stores to addresses in the log filter are not recorded. The filter is based on virtual addresses, and it is safe to clear it at any time.

Like LogTM, LogTM-SE uses a "sticky" coherence state to ensure evicted cache blocks continue to receive coherence requests. If a block is evicted, then its directory state is not cleared if it is present in a signature.

Signature-based conflict detection mechanisms can be extended to allow threads to be pre-empted mid-way through a transaction or to be rescheduled onto a different processor. To handle pre-empted threads, LogTM-SE employs a per-process summary signature. This summary signature is maintained by the OS and records the union of the read/write-sets of any pre-empted threads' suspended transactions. The summary signature must be checked for conflicts on every memory access, and so it would typically be distributed to all processors running threads from the process involved. To maintain the summary signature, if the OS deschedules a thread then it (*i*) merges the thread's signatures into the summary, (*ii*) interrupts other thread contexts running the process, and (*iii*) installs the new signature into each of these contexts. The summary signature is not checked on coherence requests but only on accesses. When the OS re-schedules a thread, it copies the thread's saved signatures from its log into the hardware read/write signatures. The summary signature itself is not recomputed until the thread commits, to ensure blocks in sticky states remain isolated after thread

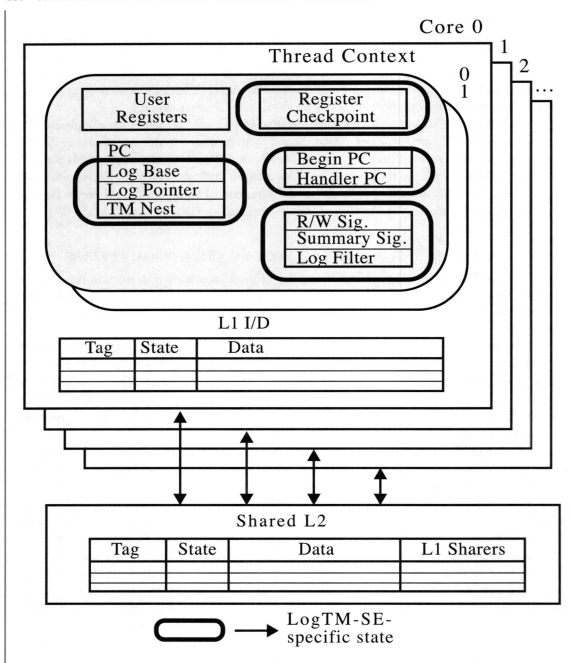

Figure 5.12: LogTM-SE hardware organization (adated from Yen *et al.* [337]).

migration. Upon commit, LogTM-SE will then trap to the OS to provide an updated summary signature to all threads.

Swift *et al.* extend this scheme and show how to include support for a virtual machine monitor (VMM), in addition to an OS [311]. Using a VMM introduces several additional problems. First, the VMM may migrate application threads transparently to the intermediate OS, losing processor-resident state in the process. This problem can be addressed by making summary-signatures VMM-private state, rather than having them be managed by the OS. The second problem is that the VMM may virtualize the memory addresses used by the OS, adding a second layer of translation below the OS's view of physical memory, and the actual addresses used by the underlying hardware. This prevents the OS from updating signatures if the VMM relocates a page in physical memory. Swift *et al.* suggest addressing this problem by using summary signatures based on *virtual* addresses, rather than physical.

5.4.2 USING PERSISTENT META-DATA

UTM. Ananian *et al.* [15] propose Unbounded Transactional Memory (UTM), an HTM scheme based on per-block memory metadata to decouple transactional state maintenance and conflict checking from hardware caches. UTM embodies two architectural extensions (Figure 5.13): a single memory-resident data structure (XSTATE) to represent the entire state of all transactions in the system and main-memory metadata extensions for each memory block in the system.

The XSTATE has a per-transaction commit record, and a linked list of log entry structures. The commit record tracks the status of the transaction (pending, committed, or aborted state), along with a timestamp which is used to resolve conflicts among transactions. Each log entry is associated with a memory block read or written in the transaction, holding a pointer to the block in memory, the old value of the block, and a pointer to the transaction's commit record. These pointers also form a linked list to all entries in the transaction log that refer to the same block. UTM uses the XSTATE to record the original value of a memory location, in case it is needed for recovery. The OS allocates the transaction log and two hardware control registers record the range for the currently active threads.

The metadata extensions consist of the addition of a RW bit and a *log pointer*. If the RW bit is not set, then no active transaction is operating on that memory block. If the bit is set, then the pointer for the memory block points to the transaction log entry for that block. This check is performed by the UTM system on all memory accesses. The access may result in a pointer traversal to locate the transaction that owns this block.

The log pointer and the RW bit for each user memory block are stored in addressable physical memory to allow the operating system to page this information. Every memory access operation (both inside and outside a transaction) must check the pointer and bits to detect any conflict with a transaction. Page tables also record information about the locations. Since, during a load or store operation, an XSTATE traversal may result in numerous additional memory accesses, the processor must either support the restart of a load or store operation in the middle of such a traversal, or

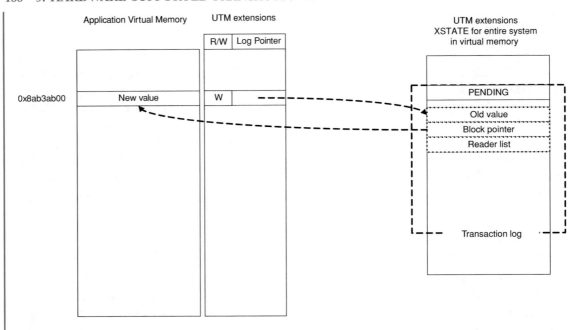

Figure 5.13: UTM data structures.

the operating system must ensure that all pages required by an XSTATE traversal are simultaneously resident. Hardware directly operates on these architectural extensions.

Conceptually, UTM does not require caches for functional correctness. However, it uses a conventional cache-based HTM for performance; and invokes the UTM machinery only when transactions overflow. The UTM data structures provide version management capabilities via hardware logging into software. To ensure correct interaction between non-overflowed HW transactions and overflowed UTM transactions, the HTM hardware must always check the log pointer and the RW bit for an accessed memory block even if no transaction overflows. Since each memory block has a pointer to the UTM transaction log, hardware can determine the state of each transaction currently accessing that memory block, and commits and aborts require atomic operations only on the transaction status in the log. However, determining all conflicting transactions requires a walk of the log pointers. Care is required when transaction logs are reused.

OneTM.   Blundell *et al.* [34] propose OneTM, an unbounded scheme similar in concept to UTM. While UTM allows multiple overflowing transactions to concurrently execute, OneTM allows only a single overflowing transaction but allows multiple normal HW transactions to execute in parallel with the single overflowing transaction. This simplification enables simpler hardware changes than UTM requires. OneTM also uses a logging scheme maintained in software for version management (as LogTM does) but does not need to maintain explicit transactional state like the XSTATE in UTM since it supports only a single overflowing transaction.

Like UTM, OneTM also uses per-memory-block metadata. It also has per-memory-block RW bits. However, instead of maintaining a pointer to a log, OneTM maintains a 14-bit transaction identifier to track the overflowed transaction (OTID). As in UTM, the metadata extensions need to be recorded with the page tables and on disk when paging. To ensure only one transaction overflows, OneTM maintains a system-wide status word STSW that indicates the identity of the current overflowed transaction (if any). These become part of the architecture. These additional 2 bytes of metadata must be queried on every memory access, as in UTM, thereby increasing the payload.

Since the OneTM meta-data is simply an extension of a memory block, updating the metadata would intuitively require write permissions to the underlying memory. However, since reads must also update the meta-data, such an operation would cause exclusive requests for reads, resulting in concurrent HW transactions aborting even if they were performing only reads. To address this issue, OneTM requires the meta-data updates to be performed if the cache has the line in an owned state. This requires the coherence protocol to support a state where, among multiple caches that share the line, at least one cache actually owns the cache line (as if it were in a dirty-shared O state). If the cache line is in S state, the cache controller must request an O state prior to updating the meta-data. This ensures a shared line's metadata update does not require an exclusive ownership request.

Another challenge is in clearing the metadata on commits and aborts. In UTM, since the meta-data had a pointer to the transaction log, the current status was always known (ignoring the corner case of log reuse). With OneTM, only the OTID is known. OneTM adopts a lazy approach to clear the metadata. A committing overflowing transaction clears the overflow bit in the STSW. This allows other transactions (which must always check this overflow bit) to ignore the metadata if there is no overflowing transaction in the system. When a transaction enters an overflow state, it also increments the OTID. This allows transactions to determine the validity of the metadata; an increment of the OTID in the STSW forces the OTIDs in memory metadata to be stale. This does introduce the issue of wrap arounds in OTIDs causing false conflicts; this is a performance issue and these conflicts eventually resolve.

As in UTM, pre-emption during a transaction requires the operating system to save all per-thread state, including the OTIDs and metadata.

Both UTM and OneTM rely on a conventional cache-based HTM to provide performance. They extend this to invoke the metadata management on overflows, and they require the processor hardware to check the meta-data on *all* accesses. They, therefore, have the hardware cost and complexity of both supporting an HTM and supporting the metadata extensions for the overflow cases.

TokenTM.   Bobba *et al.* take a fundamentally different approach to metadata management in TokenTM (Figure 5.14) [39]. Instead of using metadata only for overflows, their design uses metadata as a fundamental unit of conflict tracking and detection at all times. This enables a seamless design without special casing to distinguish overflowed from non-overflowed cases.

TokenTM does this by introducing the notion of *transactional tokens* captured in per-memory block metadata, inspired by the idea of token-based coherence mechanisms [213]. The key idea is to assign each memory block some number, T, of tokens. A transaction must acquire at least one token to read the block and must have all tokens to write the block. TokenTM introduces concepts of *metastate fission* and *fusion* to allow for concurrent updates. TokenTM uses these mechanisms as its sole conflict-detection technique. Consequently, unlike the schemes discussed previously in this section, it does not require a conventional HTM.

Conceptually, the token state for a given memory block can be represented in the form $\langle c_0, c_1, \ldots, c_i \rangle$ where c_i is the count of tokens held by a thread i. If a transaction on thread i reads the block, it requires at least one token in c_i. To write the block, the thread must obtain all tokens: it cannot write if any other c_k $(k \neq i)$ is non-zero. While, conceptually, the metadata tracks a count of tokens held by each thread for the given block, the TokenTM implementation represents the metadata as a pair (*sum, TID*), where *sum* is the total number of tokens (out of T) acquired for the block and if the sum is 1 or T, then *TID* corresponds to the owner. For example: $\langle 0, 0, 0, 0 \rangle$ is represented as $(0, -)$, $\langle 0, 0, 1, 0 \rangle$ is represented as $(1, 2)$, $\langle 0, T, 0, 0 \rangle$ is represented as $(T, 2)$, and $\langle 0, 1, 1, 1 \rangle$ is represented as $(3, -1)$.

Since the TokenTM metadata implementation is a conservative summary of the actual metadata, TokenTM also records these tokens in memory (in a manner reminiscent to memory versioning in LogTM). Each software thread maintains an in-memory list of (*addr, num*) pairs. Tokens therefore exist in two locations: as metadata bits per memory block and in the software log of each thread. For performance reasons, hardware directly operates on this log to perform efficient low-overhead appends to the log. Maintaining tokens in hardware enables fast conflict checks and maintaining them in software allows for software conflict management.

TokenTM piggy backs all token metadata manipulation operations onto existing coherence messages. If there is only a single cached copy of a given block then the owner of the data has exclusive permission to update its metadata (just as they have permission to update the associated data itself). The challenge occurs when a block is shared by more than one cache. The naïve solution of acquiring exclusive permission to update the metadata would introduce contention between concurrent readers and render the S state ineffective; TokenTM uses metadata on every memory access, and so the performance impact would be worse than with OneTM. To address this problem, TokenTM introduces the concepts of *metadata fission* and *fusion*. TokenTM allows multiple transactional readers to locally update their metadata in a local copy for shared lines. Metastate fission occurs when a shared copy of a cache block is created. Metastate fusion occurs when shared copies are merged following either a writeback or on an exclusive request. Transactional writers must request exclusive access, meaning that the token metadata that they see will have already been fused.

On commits and aborts, metadata must be returned to the memory blocks. In the event of an abort, this is done normally since the log is walked. However, on commits, these operations can

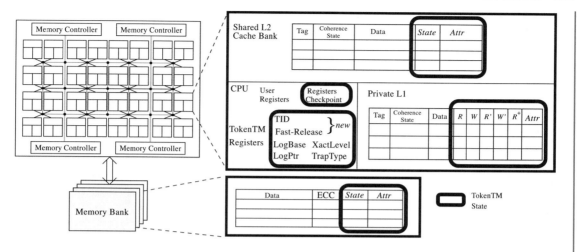

Figure 5.14: TokenTM hardware organization (adapted from Bobba *et al.* [39]).

be a significant overhead. TokenTM introduces a fast-release mechanism which resets the token log and requires additional L1 cache metabits that are flash cleared. TokenTM can also support System-V style shared memory. As in UTM and OneTM, the operating system must be aware of these extensions. This requires an instruction to free the R and W bits prior to scheduling a new thread. Paging requires the VM system to clear metastates on initialization, save them on page out, and restore them on page in. The OS must also manage the thread identifiers.

TokenTM stores 16 metabits per 64-byte memory block using ECC codes. The key idea is to change the ECC schemes to group four words together to protect 256 data bits with SECDED using 10 check bits. This makes a 22-bit code word available, which can represent 16 metabits with 6 check bits. Alternatively, explicitly reserving part of the physical memory for metabits like TokenTM for 3% more area. Baugh *et al.* also propose an implementation utilizing ECC bits [28].

VTM. In contrast to systems with per-block metadata such as UTM, OneTM and TokenTM, Rajwar *et al.*'s Virtual Transactional Memory (VTM) proposal [251] took a slightly different approach for accessing and managing metadata. Instead of extending each memory block with metadata, VTM maintains all the information for transactions that exceed hardware resources in a table in the application's virtual address space. VTM also records all tentative updates in the overflow structures while leaving the original data in place. It also requires the processor to selectively interrupt load and store operations and invoke microcode routines to check for conflicts against the overflowed transactional state in the table.

Figure 5.15 shows the VTM system organization. VTM added a microcode/firmware capability in each processor to operate on the data structures, similar to hardware page walkers in today's

Processor and Cache with support for
HW transactional memory

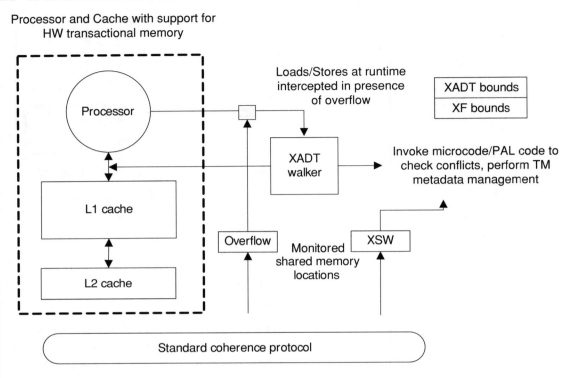

Figure 5.15: VTM hardware organization.

processors. In the absence of any overflows, VTM executed transactions using HTM without invoking any VTM-related machinery or overhead. In the presence of overflows, the VTM machinery on each processor would take appropriate actions without interfering with other processors.

The software data structure is similar in concept to the UTM overflow data structure. The XADT maintains information for transactions which have either exceeded hardware caches or have been suspended. The XADT is common to all transactions sharing the address space. Transactional state can overflow into the XADT in two ways: a running transaction may evict individual transactional cache lines or an entire transaction may be swapped out, evicting all its transactional cache lines. Each time a transaction issues a memory operation that causes a cache miss and the XADT overflow count signals an overflow in the application, it must check whether the operation conflicts with an overflowed address. In addition, each thread has a transaction status word (similar to UTM and OneTM). This transaction status word XSW is continually monitored by an executing thread.

Transactions that fit in the cache do not add state to the XADT, even if another transaction has overflowed. Processors continually monitor a shared memory location that tracks whether the XADT has overflowing transactions: XADT lookups are performed only if a transaction has overflowed.

To accelerate the conflict check and avoid an XADT lookup, VTM used a software conflict filter, the XADT filter (XF). The XF returned whether a conflict existed without requiring to look up the XADT. The XF essentially replaces the per-memory-block metadata extensions. The XF is a software filter (implemented as a counting Bloom-filter [33; 102]) that helps a transaction to determine whether a conflict for an address exists. VTM uses this filter to determine quickly if a conflict for an address exists and avoid an XADT lookup. Strong isolation ensures that the committed lines copied from the XADT to memory are made visible to other non-transactional threads in a serializable manner. Other threads (whether in a transaction or not) never observe a stale version of the logically committed but not yet physically copied lines. Conflicts between nonoverflowing hardware transactions are resolved using HTM mechanisms. For transactions that have overflowed, a processor requesting access to such an address will detect the conflict when it looks up the XADT. This localizes conflict detection, allows conflicts with swapped transactions to be detected, and avoids unnecessary interference with other processors.

VTM implements a deferred-update TM system. It leaves the original data in place and records new data in the XADT. When an overflow occurs, the VTM system moves the evicted address into a new location in the XADT. The transaction state for the overflowing transaction is split between caches and the XADT. In a deferred-update system, reads must return the last write in program order. For performance, reads must quickly know which location to access for the buffered update. The processor caches the mapping of the original virtual address to the new virtual address in a hardware translation cache, the XADC. This cache speeds subsequent accesses to the overflowed line by recording the new address of the buffered location.

VTM requires the virtual address of the evicted cache line to be available for moving it into the XADT. When a context switch occurs, VTM moves transactional state from the hardware cache to the XADT. This increases the context switch latency for such transactions.

When a transaction overflows, the Overflow monitored location changes to 1 and processor-local VTM machinery gets involved. The processor running the overflowing transaction performs overflow transaction state management. The assists add metadata information about the overflowing address into the XADT. An XADT entry records the overflowed line's virtual address, its clean and tentative value (uncommitted state), and a pointer to the XSW of the transaction to which the entry belongs. The other processors perform lookups against overflow state. A hardware walker, similar to the page miss handler, performs the lookups.

Figure 5.16 shows the coexistence of overflowed software and hardware state. The left shows the software-resident overflow state and the right shows the hardware-resident non-overflow state for a transaction. The overflow entries have a pointer to the XSW, which allows other transactions to discover information about a given transaction. The XF is a software filter (implemented as a counting bloom-filter [33; 102]) that helps a transaction to determine whether a conflict for an address exists. VTM uses this filter to determine quickly if a conflict for an address exists and avoid an XADT lookup. In the example shown, virtual addresses G and F map to the same entry in XF and thus result in a value 2 for that entry. Strong isolation ensures that the committed lines copied from the XADT to

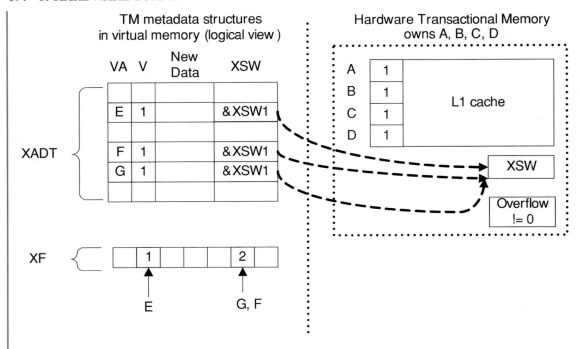

Figure 5.16: VTM hardware and software modes.

memory are made visible to other non-transactional threads in a serializable manner. Other threads (whether in a transaction or not) never observe a stale version of the logically committed but not yet physically copied lines.

5.4.3   USING PAGE TABLE EXTENSIONS

IBM 801.   An early example of using page-table extensions for transactions is the IBM 801 storage manager [57; 246]. This system provided hardware support for locking in database transactions. Specifically, the system associated a lock with every 128 bytes in each page under control of the transaction locking mechanism. It extended the page table entries (PTE) and translation look-aside buffers (TLB) to do so. If, during a transaction, a memory access did not find the associated lock in the PTE in an expected state, the hardware would automatically invoke a software function. This software function then performed the necessary transaction lock management. To accommodate this mechanism, each segment register in the 801 architecture has an S protection bit. If the S bit was 0 the conventional page protection is in effect. If the S bit was 1, the transaction locking mechanism was applicable to the segment. Figure 5.17 shows the architectural extensions; all transaction management is implemented in software, in response to invocations from the hardware.

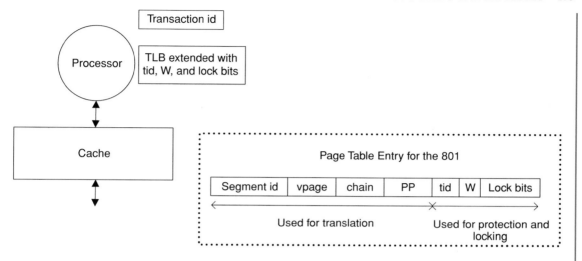

Figure 5.17: The 801 transaction locking extensions.

This implicit invocation of transaction management functions by hardware was analogous to the implicit invocation of page fault handlers by hardware to implement virtual memory. The designers of the 801 storage manager argue that removing the need to insert library calls directly into software simplifies software development and engineering—e.g., requiring explicit transaction management function calls wherever the program references transaction data presents difficulties for multiple languages. In addition, the called functions need use-specific parameters, thus complicating software engineering.

The original system was described in the context of a uniprocessor implementation accessing files on disk. Conceptually, however, the ideas are equally applicable to a multiprocessor implementation and data held in memory.

The software component of the 801 system maintains information about locks in a separate software structure called the lock table. Each entry in the lock table, called a *lockword*, has fields similar to those in the PTE. A lockword is allocated when a transaction first accesses a page following a commit, and it is freed on the next commit. Two lists are maintained to allow fast accesses to this table. One list is indexed by a transaction ID, to identify all `lockword` entries belonging to that transaction. Commit operations use this list to find all the locks held by the committing transaction in all pages. The second list is indexed by the segment id and the virtual page within the segment. The lock-fault handler uses these lists to find all locks of any transaction in the referenced page, to detect conflicts, and to remember locks granted or released. The hardware triggers a lock-fault interrupt if the locks in the PTE for a page are not those of the current transaction, and control transfers to the lock-fault handler. The handler searches the lock table and makes the transaction wait if there are conflicting locks, or grants and adds the requested lock in the table. The lock table itself is held in pageable memory and is accessed in a critical section. The table itself is similar to the ownership tables in many word-based software transactional memory systems (Section 4.5.3).

PTM.   Chuang *et al.*'s unbounded page-based transactional memory (PTM) [62] provides extensions to a processor's virtual memory mechanisms to allow transactional data to be evicted from a cache. The design assumes that short transactions are handled by an ordinary bounded-sized HTM, and so it focuses specifically on techniques for use as a fall-back in the case of overflow—e.g., context switches, processor exceptions, and the management of inter-process shared memory.

The crux of the idea is for the hardware to maintain a *shadow page* for each physical page that holds transactional data that has been evicted from a cache. A shadow page table (SPT) maps each original physical page onto the address of its associated shadow page. (The physical space itself is allocated on demand, and a similar table records mappings for pages that have been evicted from physical memory). In addition, each transaction has a list of transaction access vectors (TAVs) which identifies the SPT entries that it has accessed, and a per-cache-block bitmap, within those pages, of the locations in its read-set and write-set. These per-cache-block bitmaps enable conflict detection to be managed on a fine granularity, even though versioning is managed on a per-page basis.

Although the SPT and TAVs are conceptually held in main memory, extensions to the memory controller are responsible for caching and aggregating information from these structures so that it is not usually necessary to walk them directly.

A simpler Copy-PTM variant ensures that, for each physical page, the shadow page holds a snapshot of the current non-speculative data, while speculative updates are made eagerly to the original copy of the page. This means that, upon commit or roll-back, it is necessary to update one or other of these pages.

A more complex Select-PTM variant relaxes this condition and maintains a per-cache-block bitmap to record which of the two pages holds the non-speculative copy of the block. This avoids the need to copy data on commit or roll-back (the bitmap can be adjusted instead), but it can introduce additional work to merge the contents of the two pages if physical memory is scarce.

XTM.   Chung *et al.* examined design considerations for unbounded TMs, and they proposed a page-based *eXtended Transactional Memory* (XTM) system [64]. In the basic design, a transaction has two execution modes: all in hardware or all in software. An overflow exception is triggered when a transaction cannot be executed in hardware, and it is then aborted and re-executed in software mode. The software mode uses page faults to trap the first transactional access to each page by a given transaction, creating additional copies on-demand in virtual memory. For reads, a single *private* copy of the page is created. For writes, an additional *snapshot page* is created to hold a clean copy of the page's pre-transaction state.

The snapshot page is used for commit-time conflict detection: if it differs from the contents of the original page, then a conflict is signaled. If all of the snapshot pages are validated, then the contents of the private pages are copied back to the original locations. A sparsely-populated per-transaction page-table provides the physical location of the private pages. The software-mode commit can be made atomic by using a TCC-style global commit token or by using TLB-shootdown to ensure that the committer has exclusive access to the pages involved.

Chung *et al.* also describe extensions to allow a hardware-mode transaction to gradually overflow to virtual memory (rather than needing to be aborted) and a cache line-level variant based on tracking read/write access sets at a sub-page granularity.

5.5   EXPOSING HARDWARE MECHANISMS TO STMS

We have so far discussed HTM designs where the hardware attempts to execute the transaction. In this section, we discuss a final class of TM systems where the individual HTM hardware mechanisms are exposed directly to the STM. In these systems, the STM utilizes the HW mechanisms to improve its performance.

5.5.1   ACCELERATING SHORT TRANSACTIONS AND FILTERING REDUNDANT READS

One set of techniques for hardware-accelerated STM systems is to use hardware to perform conflict detection for short cache-resident transactions and to provide a form of conservative filtering to avoid repeated STM work if a longer transaction reads from the same location more than once.

Saha *et al.* [275], Marathe *et al.* [211], and Shriraman *et al.* [292] all propose adding additional bits per cache line and exposing these as part of the architecture. At a high level, this is reminiscent of the use of cache-based W-bits to filter repeated logging in LogTM (Section 5.3.1) and the use of software-implemented filters in Bartok-STM (Section 4.2.1).

Saha's proposal focuses on accelerating invisible-read STM systems by extending each cache line in the first level data cache with 4 *mark bits*: one mark bit per 16 byte block in a 64-byte cache line. Software uses a new load instruction to set the mark bit as part of the load operation. The proposal provides new instructions to test and clear the mark bits for a specified address and to clear the mark bits for all addresses in the data cache. The mark bits are cleared when the line is evicted or receives a coherence invalidation. A new software-readable saturating counter, the *mark counter*, tracks the number of such evictions and invalidations. This enables software to identify addresses to monitor and determine when monitoring was lost. If the mark counter is zero, then software could be sure no other thread attempted to write the locations that were marked and thereby skip the validation phase. In effect, this achieves the behavior of visible readers, without the software cost of managing shared metadata.

As in Herlihy and Moss' HTM, a loss of monitoring does not automatically result in a transaction abort; a transaction could continue executing with inconsistent data until software performed an explicit validation sequence. In an optimization, this form of TM could skip the read barrier code completely and only resort to read barrier logging if a marked line was evicted. Saha *et al.* also suggested the use of mark bits to filter write barrier operations.

The Alert On Update (AOU) proposal [211; 292] uses similar bits to those proposed by Saha *et al.*. Each cache line was extended with a bit and software could set the bits as part of a new form of load instruction, and they could clear the bits for specific addresses (or for all addresses in the data cache). However, in a key difference from Saha *et al.*'s design, the AOU approach provides a synchronous abort capability, instead of relying on software to poll for loss of monitoring, AOU triggers a software event when the loss of monitoring occurred. This requires software to register an *alert handler* using a new instruction. When the monitoring is lost (due to eviction or invalidation), the alert handler is invoked. AOU also provides the capability to mask alerts from being delivered (similar to disabling/enabling interrupts). The asynchronous abort capability enables AOU to detect abort conditions as they occur and prevents the STM from continuing to execute with inconsistent data. This mechanism could also be used to monitor status metadata in an STM, allowing immediate notification of conflicts. The proposal requires additional architectural registers to record the handlers and other miscellaneous information.

Harris *et al.* introduce a *dynamic filtering* (`dyfl`) mechanism which could be used to avoid repeated read-barrier operations in an STM system [140]. Unlike the previous proposals in this section, `dyfl` aims to be decoupled from the cache, maintaining a separate table of recently-executed barrier operations. Table entries are explicitly added and queried by the STM, and the table is explicitly flushed at the end of each transaction. The table is not modified in response to coherence traffic, and it does not aim to accelerate commit operations. However, the design aims to support a series of additional applications beyond STM—e.g., read and write barriers for garbage collection.

5.5.2   SOFTWARE CONTROLLED CACHE COHERENCE

Both Saha *et al.*'s proposal and AOU leave the existing cache coherence protocols unchanged. They, therefore, result in conflicts occurring when the coherence protocol makes conflicting requests (irrespective of when the conflict would actually be detected by an STM system).

To decouple conflict detection completely from the cache coherence protocol, the RTM [211; 289] and FlexTM [290; 291] systems also proposed Programmable Data Isolation (PDI). As outlined in Section 5.3.4, with PDI, software has explicit control over when an address in the data cache participates in the cache coherence protocol. Software uses two new instructions, TLoad and TStore, for doing so. TStore operations use the data cache to buffer transactional writes (and use lower levels of memory to save old copies of the line, the same way as most HTM systems) but tag the line in such a way that it does not participate in cache coherence. This incoherence allows lazy detection of read/write and write/write conflicts. PDI introduces two new stable states to the cache coherence protocol: TI and TMI (goes to M on commits and I on aborts). T implies *threatened* and is a new indication provided to a TLoad operation that the line may be in TMI state somewhere in the system and therefore may be a potential conflicting writer. Similarly M, E, and S state lines that are tagged by TLoad operations transition to TI when written by other processors, but continue to be readable even though they are incoherent. Software has to ensure correctness by handling incoherence properly. Commit operations are done through a CAS-commit instruction. PDI requires extensive support in the cache coherence protocol to support incoherence. The hardware itself does not provide any transactional execution properties—it is up to the associated STM to use the hardware mechanisms correctly.

5.5.3   EXPOSED SIGNATURES TO STMS

SigTM [48] proposes a hardware acceleration based on signatures instead of software-maintained read/write sets or the data cache. SigTM uses hardware signatures to track read- and write-sets and perform conflict detection. However, software continues to implement all remaining actions such as access identification and version management. Software version management implies a write-set must also be maintained in software to allow recovery.

SigTM has three key relevant actions: `SigTMtxStart` (take a checkpoint and enable read-set signature lookups for exclusive coherence requests), `SigTMWriteBarrier` (add address into the write-set signature and update the software write-set), and `SigTMreadBarrier` (check if the address is in the software write-set; if not, add into the read-set). If an exclusive coherence request hits the read set signatures, a conflict is signaled and the transaction aborts. Aborts result in a signature reset. Write-sets are not looked up until the commit stage—this allows for write-write concurrency.

The commit sequence is as follows. First, coherence lookups are enabled in the write-set signatures for all incoming coherence requests. Then, the write-set is scanned and exclusive access requests are issued for every address in the set using a new `fetchex` (fetch-exclusive) instruction. Any request that hits a write-set is NACKed. As a result, a `fetchex` instruction may time out and invoke a software handler. Next SigTM scans the write set and updates memory. Performing lookups to read-set signatures results in the transaction with visible reads, thereby avoiding the inconsistent read problems with invisible read STMs. Since the read set is exposed to all coherence protocol requests, the STM can naturally achieve strong isolation.

Decoupling access tracking from caches introduces complications in dealing with lines that have been evicted: the processor may not receive subsequent coherence requests and therefore will be unable to perform lookups to its signature. This is a situation similarly faced with LogTM, LogTM-SE and FlexTM. SigTM assumes a broadcast protocol where all processors see all requests, irrespective of whether their cache hierarchy has a copy of the block. However, this is not how many modern multicore systems are implemented because they often use their cache hierarchies as filters to reduce the coherence traffic that they must see. Regardless, using signatures requires them to be exposed to all traffic to ensure no address lookup is missed.

These signatures are software-readable and can be updated. They face similar challenges to those faced by LogTM-SE when it comes to thread de-scheduling, thread migration, conflict checking during suspension, and page remapping. Approaches similar to LogTM-SE may be used to address these challenges.

A key distinction between SigTM and the earlier hardware-accelerated STMs is that SigTM cannot execute without hardware support for signatures whereas hardware-accelerated STMs and most Hybrid TMs can execute completely in software without hardware support. This is, however, primarily due to the goal of SigTM to achieve low overhead strong isolation.

5.5.4   EXPOSING METADATA TO STMS

Baugh *et al.* [28] propose exposing memory metadata explicitly to software. This enables an STM in a hybrid system to use this metadata for conflict checking and to achieve strong isolation. The paper proposes the concept of a User Fault On (UFO) mechanism for each block in the memory system using new bits called UFO bits. These bits serve to protect access to the block.

When a SW transaction operates on a memory location, it sets the appropriate read and/or write UFO bits to write-protect values read or read/write-protect values that they write. While executing a SW transaction, a thread disables detection of these protection violations for its own accesses. Consequently, threads using SW transactions can access memory as normal (using STM for their synchronization), but threads accessing the locations nontransactionally or via HW transactions will receive protection violations. This allows a SW transaction to achieve strong isolation with respect to other HW transactions and non-transactional accesses without requiring the hardware to check the STM metadata.

Baugh *et al.* describe an example STM implementation using an ownership table. To provide strong isolation, requires installing memory protection for the transactionally cached blocks whenever the otable entries are created/upgraded. A fault-on-write bit is installed on the USTM read barriers and a fault-on-read and fault-on-write is installed on USTM write barriers. USTM transactions themselves disable UFO faults. Operations to the USTM structures and UFO bits use a lock to ensure no data races. The operations add minimal overhead to the underlying STM. When a conflict does occur (i.e., the HTM or an ordinary transaction accesses the cache and finds an incompatible UFO bit set), then the faulting thread vectors to a fault handler that was registered by the STM.

Since the UFO bits are required to be coherent, they may result in aborting the HW transactions even on reads. However, the paper does not find this to be a major issue. While this is a different observation from the TokenTM proposal, the use of metadata is distinct. The UFO proposal utilizes these metadata only when a software transaction is executing whereas TokenTM uses them all the time.

Figure 5.18 describes a proposed UFO implementation. UFO adds two bits of information to every cache block of data, extending to all levels of the virtual memory hierarchy. Multithreaded processors where multiple threads share the same data cache require only one set of UFO bits. Existing cache coherence ensures these UFO bits are consistent across all caches.

The paper proposes an implementation utilizing ECC bits. The idea is to encode ECC at a larger granularity (similar to TokenTM) and reuse the resulting free ECC bits. This requires support in the memory controllers to re-encode ECC and to interpret the remaining ECC bits as UFO bits. The operating system must save and restore these UFO bits when physical pages swap to and from disk.

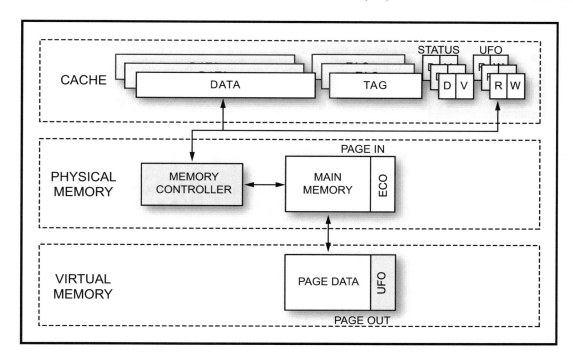

Figure 5.18: UFO metadata implementation (adapted from Baugh *et al.* [28]).

5.6 EXTENDING HTM: NESTING, IO, AND SYNCHRONIZATION

In this section, we examine a number of extensions to HTM systems, focusing on three particular directions: supporting different forms of nesting between transactions, supporting IO operations within transactions, and supporting synchronization between transactions running on different processors. The mechanisms used for these three extensions are often related to one another, and so we consider the extensions as a whole, rather than treating each separately.

Moravan *et al.* [228] and McDonald *et al.* [215] developed mechanisms to accommodate closed nesting and open nesting in HTM. Both techniques conceptually divide a transactions read/write sets into a series of segments for each nesting level. Closed nesting is supported by allowing the innermost segment to be aborted without aborting the complete transaction. Open nesting is supported by allowing the innermost segment to be committed directly to memory.

In systems using cache-based *R/W*-bits for tracking accesses, one approach is to replicate these bits and provide separate pairs for each nesting level [215; 228]. McDonald *et al.* propose a second implementation in which tentative state is held in different lines in the same cache set, as sketched by Moss and Hosking [229]. Flattening can be used to optimistically implement closed nesting once these bits are exhausted, or hardware-software hybrids can be used.

Similarly, Bulk supports a closed-nested model. A separate read and write signature is maintained for each nested level. Incoming write signatures and addresses are checked for conflicts (via intersection or membership) with each of these nested signatures. If any intersection is found, then execution rolls back to the beginning of the aborting transaction. When the outer transaction commits, then a union of all the write signatures is broadcast to other processors for conflict detection. Bulk can track multiple read and write-sets for the nesting levels without requiring per-nesting level cache bits. The paper suggests moving signatures into a memory location if more nesting levels are required than are provided by the processor.

In addition to ordinary nesting, Moravan *et al.*'s design allows a transaction to register *commit actions* and *compensating actions* and to execute *escape actions* during which memory accesses are made directly (with conflict detection disabled). These actions are intended for low-level interfacing, e.g., for use in wrappers around system calls performing IO. McDonald *et al.*'s design provides facilities for two-phase commit (2PC) using commit-time callbacks and for software violation-notification handlers to be invoked when a conflict occurs. To simplify the implementation of 2PC, the hardware effectively votes last, so that it can immediately commit or roll-back its updates.

As in software implementations, programmer care is required to ensure that nesting and non-transactional operations are used in a disciplined way. Particular care is needed if locations might be shared between a parent and an open-nested child—and hardware implementations can add the complexity that such sharing may need to be determined at a per-cache-line granularity, rather than in terms of precise addresses. McDonald *et al.* [215] introduce rules for safe use of these operations.

Lev and Maessen proposed an alternative "split transaction" technique for implementing nesting in hardware, along with extensions such as commit actions and 2PC [190]. Lev and Maessen's approach assumes only simple non-nesting transactions in the hardware, and it builds richer semantics for transactions over this by tracking explicit read and write-sets in software. In effect, each hardware transaction provides an atomic update to these read and write-sets. For instance, on entering a closed nested transaction, a new hardware transaction is started which re-reads the enclosing transaction's read set (so that conflicts are detected on it) before updating the read and write sets with the enclosed transaction's memory accesses. Commits are performed by writing back the contents of the write set. This approach is flexible, and it avoids the need to support nesting explicitly in hardware. However, it does incur the cost of maintaining read and write-sets in main memory, it requires software to be compiled to operate on these sets, and it does not (in itself) support transactions that are longer than those supported by the underlying HTM.

Baugh and Zilles [346] introduced a `pause` operation, allowing a section of code to execute non-transactionally while a transaction remains active on the current processor. The non-transactional code is intended to manage thread-local information (e.g., a thread-local allocator), invoke system calls in a controlled way, manage lists of abort-handlers and commit-handlers, and so on. In addition, Baugh and Zilles describe a mechanism to allow a thread running a hardware transaction to block, using a `retry`-like operation, in which the thread links itself to data structures whose updates should trigger it to be woken (as in the software implementation in Section 4.6.2), before requesting that the OS reduce the thread's priority to the minimum possible; its transaction will be aborted (and the thread woken) when a conflicting update is made. A similar mechanism can be used to allow a thread running one transaction TA to block until a conflicting transaction TB has completed: TA updates a per-transaction metadata record for TB, and a new form of processor exception is raised to notify the OS scheduler. When TB commits, it proceeds to request that the thread running TA be rescheduled. The use of transactions helps avoid race conditions in managing these operations.

Hofmann *et al.* argue that a small number of hardware primitives can be used in place of direct support for open nesting (or for unbounded transactions) [158; 159]. They describe their experience using an HTM system to implement synchronization in the Linux kernel, extending a bounded-size HTM with support for non-transactional operations, for blocking, and for non-transaction interaction with contention management. Their design for non-transactional operations is similar to Baugh and Zilles's `pause` operations, except that while Hofmann *et al.* do not allow these operations to access transactional data, they *do* allow them to use transactions themselves. Blocking is performed by an `xwait` operation, which waits for an address to hold a specific value before adding the address to the transaction's read set. This allows a transaction to probe an address (e.g., a status field, or a lock's implementation), without inducing a conflict. Finally, an `xcas` operation is used in non-transactional code that is integrated with contention-management. This is used to avoid some of the pathologies discussed by Volos *et al.* [321] in managing shared transactional-non-transactional data structures.

5.7   SUMMARY

We have discussed the main directions explored in the design and implementation of HTM systems. As we have seen, the HTM design is often directly impacted by the targeted software model and expectations of performance and software ecosystem impact. HTMs that aim to minimize software impact require more complex mechanisms than hardware geared primarily toward accelerating STMs. This is analogous to hardware support for virtual memory where significant effort went into providing high performance and transparency to application software.

Hardware support has unique challenges not faced by software systems. Any hardware support becomes legacy when it ships, and it often must be supported going forward. This places great focus on correct and concise instruction set definitions, and ensuring the behavior is well defined in the presence of arbitrary software usages and interactions with other features such as virtualization, varied system software, and other competing usage models. This also directly affects validation and verification costs. Hardware design goes through a rigorous verification and validation process, and open-ended instruction set definitions greatly increase these costs. Modern microprocessors are highly complex designs that are efficient in processing a continuous stream of instructions with high throughput and parallelism; the actual execution flow through the pipelines often is different from the software view (which only observes the retirement order). New instruction set extensions to support TM must lend themselves to seamless integration into such processors; operations that interrupt execution flow can negatively impact performance. Further, cache coherence and memory ordering implementations are very complex, and significant effort goes into their validation. These challenges are especially true with extensions that provide software direct control over individual hardware mechanisms.

Unbounded HTMs that rely on metadata require such support to be architectural, much in the same way virtual memory page tables are architectural. Such support is expensive and expansive since it requires both hardware and system software changes. Further, it becomes important to get the definition correct, otherwise it becomes a legacy burden. While such HTM systems can significantly reduce the larger software ecosystem impact, they require complex changes to hardware implementations.

TM is still an open research topic, and it is as yet unclear what the right hardware model would be. This is also affected by the realities of software deployment, the extent of software analysis and recompilation expected and accepted, and the expectation of legacy support. As the field matures, clear directions may emerge. Nevertheless, research into HTMs has developed a rich set of options and alternatives to provide TM support. An ongoing research question is exactly where the boundary should be placed between functionality provided by a hardware implementation of TM and a software system.

CHAPTER 6

Conclusions

Transactional memory (TM) is a popular research topic. The advent of multicore computers and the absence of an effective and simple programming model for parallel software have encouraged a belief (or hope) that transactions are an appropriate mechanism for structuring concurrent computation and facilitating parallel programming. The underpinning of this belief is the success of transactions as the fundamental programming abstraction for databases, rather than practical experience applying transactions to general parallel programming problems. Confirmation of this belief can come only with experience. The research community has started exploring the semantics and implementation of transactional memory, to understand how this abstraction can integrate into existing practice, to gain experience programming with transactions, and to build implementations that perform well.

This book surveys the published research in this area, as of early 2010. As is normal with research, each effort solves a specific problem and starts from specific, though often narrow, assumptions about how and where a problem arises. Software transactional memory (STM) papers offer a very different perspective and start from very different assumptions than hardware transactional memory (HTM) papers. Nevertheless, there are common themes and problems that run through all of this research. In this survey, we have tried to extract the key ideas from each paper and to describe a paper's contribution in a uniform context. In addition, we have tried to unify the terminology, which naturally differs among papers from the database, computer architecture, and programming language communities.

Looking across the papers in this book, a number of common themes emerge.

First, transactional memory is a new programming abstraction. The idea evolved from initial proposals for libraries and instructions to perform atomic multiword updates to modern language constructs such as `atomic{ }`, which allow arbitrary code to execute transactions and which often allow transactions to execute arbitrary code. Moreover, researchers have elaborated the basic concept of a transaction with mechanisms from conditional critical sections such as `retry` and mechanisms to compose transactions such as `orElse`. These language features lift a programmer above the low-level mechanisms that implement TM, much as objects and methods build on subroutine call instructions and stacks.

Control over the semantics and evolution of programming abstractions belongs in large measure to programming language designers and implementers. They need to invent new constructs and precisely define their semantics; integrate transactional features into existing languages; implement the language, compiler, and run-time support for transactions; and work with library and application developers to ensure that transactions coexist and eventually support the rich environment of modern programming environments.

This level of abstraction is what we loosely refer to as the semantic layer of transactional memory. Programmers will think at this level and write their applications with these abstractions. A key challenge is to ensure that these constructs smoothly coexist with legacy code written, and perhaps even compiled, long before TM. This code is valuable, and for the most part, it will not be rewritten. Building a system entirely around transactions is a long-term endeavor. Transactions must coexist with non-transactional code for a long time. It is also important that transactional memory systems provide features and programming conventions suitable for multiple languages and compilers, so a construct written in one language and compiled with one compiler interoperates with code written in other languages and compiled with other compilers. A further challenge is to learn how to program with transactions and to teach the large community of programmers, students, and professors a new programming abstraction.

The second theme is that, high-performance software implementations of transactional memory will play an important role in the evolution of transactional memory, even as hardware support becomes available. STM offers several compelling advantages. Because of their malleability, STM systems are well suited to understanding, measuring, and developing new programming abstractions and experimenting with new ideas. Until we measure and simulate "real" systems and applications built around transactions, it will be difficult to make the engineering trade offs necessary to build complex HTM support into processors. Changing a processor architecture has costs in engineering, design, and validation. Further, instruction-set extensions need to be supported for an extended period, often the lifetime of an architecture.

Moreover, since STM systems run on legacy hardware, they provide a crucial bridge that allows programmers to write applications using TM long before hardware support or HTM systems are widely available. If STM systems do not achieve an acceptable and usable level of performance, most programmers will wait until computers with HTM support are widely available before revamping their applications and changing their programming practice to take advantage of the new programming model.

STM and HTM complement each other more than they compete. It is easy to view these two types of implementations as competitors since they both solve the same problem by providing operations to implement transactional memory. Moreover, the respective research areas approach problems from opposite sides of the hardware–software interface and often appear to be moving in opposite directions. However, each technique has strengths that complement the weaknesses of the other. Together they are likely to lead to an effective TM system that neither could achieve on its own.

A key advantage of STM systems is their flexibility and ability to adapt readily to new algorithms, heuristics, mechanisms, and constructs. The systems are still small and simple (though they are often integrated into more complex platforms, such as Java or .NET runtime systems), which allows easy experimentation and rapid evolution. In addition, manipulating TM state in software facilitates integration with garbage collection, permits long-running, effectively unbounded transactions, and allows rich transaction nesting semantics, sophisticated contention management policies, natural exceptions handling, etc. The tight integration of STMs with languages and applications allows cross-module optimization, by programmers and compilers, to reduce the cost of a transaction.

STMs also have limitations. When STMs manipulate metadata to track read and write-sets and provision for roll-back, they execute additional instructions, which increases the overhead in the memory system and instruction execution. Executing additional instructions also increases power consumption. Languages such as C/C++ offer few safety guarantees, and thus they constrain compiler analysis and limit STM implementation choices in ways that may lead to lower performance or difficult-to-find bugs. STMs also face difficult challenges in dealing with legacy code, third-party libraries, and calls to functions compiled outside of the STM.

HTM systems have different areas of strength. A key characteristic of most HTM systems is that they are decoupled from an application. This allows an HTM to avoid the code bloat necessary for STMs. Consequently, HTM systems can execute some transactions (those that fit hardware buffers) with no more performance or instruction execution overhead than that caused by sequential execution. In addition, most HTM systems support strong isolation. Furthermore, HTM systems are well suited to low-level systems code, whose fixed data layouts and unrestricted pointers constrain STM systems. HTM systems can also accommodate transactions that invoke legacy libraries, third-party libraries, and functions not specially compiled for TM.

However, HTM systems have limitations. Hardware buffering is limited, which forces HTM systems to take special action when a transaction overflows hardware limits. Some HTM systems spill state into lower levels of the memory hierarchy, while others use STM techniques and spill to or log to software-resident TM metadata structures. Since hardware partially implements the mechanisms, it does not always offer the complete flexibility of an STM system. Early HTM systems implemented contention management in hardware, which required simple policies. Recent HTM systems have moved contention management into software. Finally, HTM systems have no visibility into an application. Optimization belongs to a compiler and is possible only through the narrow window of instruction variants.

While STMs and HTM systems both have advantages and limitations, the strength of an STM is often a limitation in an HTM, and an HTM often excels at tasks STMs find difficult to perform efficiently. We see four obvious ways in which these two fields can come together.

The first possibility is that as experience building and optimizing STM systems grows, consensus may emerge on instruction set extensions and hardware mechanisms to accelerate STM systems. Obvious areas are the high cost of tracking read, write, and undo sets and detecting conflicts between transactions. Such support must be general enough to aid a wide variety of STM systems. Many of the hardware mechanisms identified in recent proposals (Section 5.2.3) retain much of the complexity of a typical HTM but provide finer control over the hardware. The challenge here is to design mechanisms that integrate into complex, modern processors and provide architecturally scalable performance over time. Other approaches are necessary to give STM the same transparency that HTM can achieve for strong isolation, legacy code, third-party libraries, and perhaps for unsafe code.

The second possibility is for the hardware community to continue to develop self-contained HTM systems, which rely on software to handle overflows of hardware structures and to implement policy decisions. This approach preserves the speed and transparency of HTM, but it may not have the flexibility of a software system, unless the interfaces are well designed. Experimentation with STM systems may help to divide the responsibilities appropriately between software and hardware and to identify policies that perform well in a wide variety of situations and that can be efficiently supported by hardware. The challenges are to ensure that aspects of the TM system that are not yet well understood are not cast into hardware prematurely and to ensure that the system can be integrated into a rich and evolving software environment.

The third approach is to combine the strengths of the two approaches into a hybrid hardware–software TM system that offers low overhead, good transparency, and flexible policy. The contributions of the STM community can include software definition of the metadata structures and their operations, software implementation of policy, and close integration into compilers and run-time. The HTM community can contribute strong isolation, support for legacy software, TM-unaware code, and low-performance penalties and overheads.

The final possibility is that both forms of TM may be appropriate for different kinds of usage: HTM systems supporting 2–4-location atomic updates are valuable for building low-level libraries, operating systems code, and language runtime systems. However, instead of using these short transactions to implement `atomic` blocks in a high-level programming language, they might just be used to build parts of the `CommitTx` implementation, or they might be used in combination with additional hardware support for `atomic` blocks. Furthermore, even if compile-time optimizations and hardware acceleration allow `atomic` blocks' performance to be acceptable for high-level programmers, it may still be valuable to have streamlined support for short transactions entirely in hardware so that these features are available for use in settings where a full STM implementation is unsuitable (e.g., some low-level systems code).

TM holds promise for simplifying the development of parallel software as compared to conventional lock-based approaches. It is yet unclear how successful it will be. Success depends on surmounting a number of difficult and interesting technical challenges, starting with providing a pervasive, efficient, well-performing, flexible, and seamless transactional memory system that can be integrated into existing execution environments. It is not clear what such a system would look like, but this book describes a fountain of research that may provide some of the ideas.

Bibliography

[1] Martín Abadi, Andrew Birrell, Tim Harris, Johnson Hsieh, and Michael Isard. Dynamic separation for transactional memory. Technical Report MSR-TR-2008-43, Microsoft Research, March 2008. 40, 95

[2] Martín Abadi, Andrew Birrell, Tim Harris, Johnson Hsieh, and Michael Isard. Implementation and use of transactional memory with dynamic separation. In *CC '09: Proc. International Conference on Compiler Construction*, pages 63–77, March 2009. DOI: 10.1007/978-3-642-00722-4_6 40, 95

[3] Martín Abadi, Andrew Birrell, Tim Harris, and Michael Isard. Semantics of transactional memory and automatic mutual exclusion. In *POPL '08: Proc. 35th Annual ACM SIGPLAN-SIGACT Symposium on Principles of Programming Languages*, pages 63–74, January 2008. DOI: 10.1145/1328438.1328449 23, 31, 38, 40, 41, 68, 77, 145

[4] Martín Abadi and Tim Harris. Perspectives on transactional memory (invited paper). In *CONCUR '09: Proc. 20th Confernece on Concurrency Theory*, September 2009. 23

[5] Martín Abadi, Tim Harris, and Mojtaba Mehrara. Transactional memory with strong atomicity using off-the-shelf memory protection hardware. In *PPoPP '09: Proc. 14th ACM SIGPLAN Symposium on Principles and Practice of Parallel Programming*, pages 185–196, February 2009. DOI: 10.1145/1504176.1504203 35, 71, 107, 109

[6] Martín Abadi, Tim Harris, and Katherine Moore. A model of dynamic separation for transactional memory. In *CONCUR '08: Proc. 19th Confernece on Concurrency Theory*, pages 6–20, August 2008. DOI: 10.1007/978-3-540-85361-9_5 23, 95, 145

[7] Ali-Reza Adl-Tabatabai, Brian T. Lewis, Vijay Menon, Brian R. Murphy, Bratin Saha, and Tatiana Shpeisman. Compiler and runtime support for efficient software transactional memory. In *PLDI '06: Proc. 2006 ACM SIGPLAN Conference on Programming Language Design and Implementation*, pages 26–37, June 2006. DOI: 10.1145/1133981.1133985 73, 102, 109

[8] Ali-Reza Adl-Tabatabai, Victor Luchangco, Virendra J. Marathe, Mark Moir, Ravi Narayanaswamy, Yang Ni, Dan Nussbaum, Xinmin Tian, Adam Welc, and Peng Wu. Exceptions and transactions in C++. In *HotPar '09: Proc. 1st Workshop on Hot Topics in Parallelism*, March 2009. 62, 75, 80

[9] Sarita V. Adve and Kourosh Gharachorloo. Shared memory consistency models: a tutorial. *IEEE Computer*, 29(12):66–76, December 1996. DOI: 10.1109/2.546611 103

[10] Sarita V. Adve and Mark D. Hill. Weak ordering – a new definition. In *ISCA '90: Proc. 17th Annual International Symposium on Computer Architecture*, pages 2–14, May 1990. DOI: 10.1145/325164.325100 39

[11] S.V. Adve and J.K. Aggarwal. A unified formalization of four shared-memory models. *IEEE Transactions on Parallel and Distributed Systems*, 4(6):613–624, 1993. DOI: 10.1109/71.242161 143

[12] Ole Agesen, David Detlefs, Alex Garthwaite, Ross Knippel, Y. S. Ramakrishna, and Derek White. An efficient meta-lock for implementing ubiquitous synchronization. In *OOPSLA '99: Proc. 14th ACM SIGPLAN Conference on Object-Oriented Programming, Systems, Languages, and Applications*, pages 207–222, November 1999. DOI: 10.1145/320384.320402 106

[13] Kunal Agrawal, Jeremy T. Fineman, and Jim Sukha. Nested parallelism in transactional memory. In *PPoPP '08: Proc. 13th ACM SIGPLAN Symposium on Principles and Practice of Parallel Programming*, pages 163–174, February 2008. An earlier version appeared at *TRANSACT '07*. DOI: 10.1145/1345206.1345232 44

[14] Kunal Agrawal, I-Ting Angelina Lee, and Jim Sukha. Safe open-nested transactions through ownership. In *PPoPP '09: Proc. 14th ACM SIGPLAN Symposium on Principles and Practice of Parallel Programming*, pages 151–162, February 2009. DOI: 10.1145/1378533.1378553 57

[15] C. Scott Ananian, Krste Asanović, Bradley C. Kuszmaul, Charles E. Leiserson, and Sean Lie. Unbounded transactional memory. In *HPCA '05: Proc. 11th International Symposium on High-Performance Computer Architecture*, pages 316–327, February 2005. DOI: 10.1109/HPCA.2005.41 162, 164, 187

[16] Nikos Anastopoulos, Konstantinos Nikas, Georgios Goumas, and Nectarios Koziris. Early experiences on accelerating Dijkstra's algorithm using transactional memory. In *MTAAP '09: Proc 3rd Workshop on Multithreaded Architectures and Applications*, May 2009. DOI: 10.1109/IPDPS.2009.5161103 91

[17] Mohammad Ansari, Kim Jarvis, Christos Kotselidis, Mikel Luján, Chris Kirkham, and Ian Watson. Profiling transactional memory applications. In *PDP '09: Proc. 17th Euromicro International Conference on Parallel, Distributed, and Network-based Processing*, pages 11–20, February 2009. DOI: 10.1109/PDP.2009.35 90

[18] Mohammad Ansari, Christos Kotselidis, Kim Jarvis, Mikel Luján, Chris Kirkham, and Ian Watson. Advanced concurrency control for transactional memory using transaction commit rate. In *EUROPAR '08: Proc. 14th European Conference on Parallel Processing*, pages 719–728, August 2008. Springer-Verlag Lecture Notes in Computer Science volume 5168. DOI: 10.1007/978-3-540-85451-7_77 54

[19] Mohammad Ansari, Christos Kotselidis, Kim Jarvis, Mikel Lujan, Chris Kirkham, and Ian Watson. Experiences using adaptive concurrency in transactional memory with Lee's routing algorithm (poster). In *PPoPP '08: Proc. 13th ACM SIGPLAN Symposium on Principles and Practice of Parallel Programming*, pages 261–262, February 2008. DOI: 10.1145/1345206.1345246 54

[20] Mohammad Ansari, Christos Kotselidis, Mikel Luján, Chris Kirkham, and Ian Watson. Investigating contention management for complex transactional memory benchmarks. In *MULTIPROG '09: Proc. 2nd Workshop on Programmability Issues for Multi-Core Computers*, January 2009. 54

[21] Mohammad Ansari, Mikel Luján, Christos Kotselidis, Kim Jarvis, Chris Kirkham, and Ian Watson. Steal-on-abort: improving transactional memory performance through dynamic transaction reordering. In *HIPEAC '09: Proc. 4th International Conference on High Performance and Embedded Architectures and Compilers*, pages 4–18, January 2009. Springer-Verlag Lecture Notes in Computer Science volume 5409. DOI: 10.1007/978-3-540-92990-1_3 54

[22] Nimar S. Arora, Robert D. Blumofe, and C. Greg Plaxton. Thread scheduling for multi-programmed multiprocessors. In *SPAA '98: Proc. 10th Symposium on Parallel Algorithms and Architectures*, pages 119–129, June 1998. DOI: 10.1145/277651.277678 2

[23] Hagit Attiya, Eshcar Hillel, and Alessia Milani. Inherent limitations on disjoint-access parallel implementations of transactional memory. In *SPAA '09: Proc. 21st Symposium on Parallelism in Algorithms and Architectures*, pages 69–78, August 2009. An earlier version appeared at *TRANSACT '09*. DOI: 10.1145/1583991.1584015 48

[24] Hillel Avni and Nir Shavit. Maintaining consistent transactional states without a global clock. In *SIROCCO '08: Proc. 15th International Colloquium on Structural Information and Communication Complexity*, pages 131–140, June 2008. Springer-Verlag Lecture Notes in Computer Science volume 5058. DOI: 10.1007/978-3-540-69355-0_12 122

[25] Woongki Baek, Chi Cao Minh, Martin Trautmann, Christos Kozyrakis, and Kunle Olukotun. The OpenTM transactional application programming interface. In *PACT '07: Proc. 16th International Conference on Parallel Architecture and Compilation Techniques*, pages 376–387, September 2007. DOI: 10.1109/PACT.2007.74 44

[26] Tongxin Bai, Xipeng Shen, Chengliang Zhang, William N. Scherer III, Chen Ding, and Michael L. Scott. A key-based adaptive transactional memory executor. In *NSF Next Generation Software Program Workshop, held in conjunction with IPDPS*, 2007. Also available as TR 909, Department of Computer Science, University of Rochester, December 2006. DOI: 10.1109/IPDPS.2007.370498 53

[27] Alexandro Baldassin and Sebastian Burckhardt. Lightweight software transactions for games. In *HotPar '09: Proc. 1st Workshop on Hot Topics in Parallelism*, March 2009. 92

[28] Lee Baugh, Naveen Neelakantam, and Craig Zilles. Using hardware memory protection to build a high-performance, strongly atomic hybrid transactional memory. In *ISCA '08: Proc. 35th Annual International Symposium on Computer Architecture*, pages 115–126, June 2008. DOI: 10.1145/1394608.1382132 71, 107, 166, 191, 200, 201

[29] Lee Baugh and Craig Zilles. An analysis of I/O and syscalls in critical sections and their implications for transactional memory. In *TRANSACT '07: 2nd Workshop on Transactional Computing*, August 2007. DOI: 10.1109/ISPASS.2008.4510738 88

[30] Emery D. Berger, Ting Yang, Tongping Liu, and Gene Novark. Grace: safe multithreaded programming for C/C++. In *OOPSLA '09: Porc. 24th ACM SIGPLAN Conference on Object-Oriented Programming Systems Languages and Applications*, pages 81–96, October 2009. DOI: 10.1145/1640089.1640096 55, 99, 128

[31] Philip A. Bernstein. Transaction processing monitors. *Communications of the ACM*, 33(11):75–86, 1990. DOI: 10.1145/92755.92767 5, 88

[32] Philip A. Bernstein, Vassos Hadzilacos, and Nathan Goodman. *Concurrency Control and Recovery in Database Systems*. Addison-Wesley, 1987. 158

[33] Burton H. Bloom. Space/time trade-offs in hash coding with allowable errors. *Communications of the ACM*, 13(7):422–426, 1970. DOI: 10.1145/362686.362692 107, 124, 193

[34] Colin Blundell, Joe Devietti, E. Christopher Lewis, and Milo M. K. Martin. Making the fast case common and the uncommon case simple in unbounded transactional memory. *SIGARCH Computer Architecture News*, 35(2):24–34, 2007. DOI: 10.1145/1273440.1250667 150, 172, 173, 188

[35] Colin Blundell, E. Christopher Lewis, and Milo M. K. Martin. Deconstructing transactions: The subtleties of atomicity. In *WDDD '05: Proc. 4th Annual Workshop on Duplicating, Deconstructing, and Debunking*, June 2005. 30, 31, 64

[36] Colin Blundell, E. Christopher Lewis, and Milo M. K. Martin. Subtleties of transactional memory atomicity semantics. *Computer Architecture Letters*, 5(2), November 2006. DOI: 10.1109/L-CA.2006.18 30, 31, 64

[37] Colin Blundell, E. Christopher Lewis, and Milo M. K. Martin. Unrestricted transactional memory: Supporting I/O and system calls within transactions. Technical Report CIS-06-09, Department of Computer and Information Science, University of Pennsylvania, April 2006. 21, 81, 87, 141

[38] Colin Blundell, Milo M.K. Martin, and Thomas F. Wenisch. InvisiFence: performance-transparent memory ordering in conventional multiprocessors. In *ISCA '09: Proc. 36th annual International Symposium on Computer Architecture*, pages 233–244, July 2009. DOI: 10.1145/1555754.1555785 150, 154

[39] Jayaram Bobba, Neelam Goyal, Mark D. Hill, Michael M. Swift, and David A. Wood. To-kenTM: Efficient execution of large transactions with hardware transactional memory. In *ISCA '08: Proc. 35th Annual International Symposium on Computer Architecture*, pages 127–138, June 2008. 189, 191

[40] Jayaram Bobba, Kevin E. Moore, Luke Yen, Haris Volos, Mark D. Hill, Michael M. Swift, and David A. Wood. Performance pathologies in hardware transactional memory. In *ISCA '07: Proc. 34th Annual International Symposium on Computer Architecture*, pages 81–91, June 2007. DOI: 10.1145/1250662.1250674 14, 50

[41] Robert L. Bocchino, Vikram S. Adve, and Bradford L. Chamberlain. Software transactional memory for large scale clusters. In *PPoPP '08: Proc. 13th ACM SIGPLAN Symposium on Principles and Practice of Parallel Programming*, pages 247–258, February 2008. DOI: 10.1145/1345206.1345242 144

[42] Hans-J. Boehm. Transactional memory should be an implementation technique, not a programming interface. In *HotPar '09: Proc. 1st Workshop on Hot Topics in Parallelism*, March 2009. 96

[43] Hans-J. Boehm and Sarita V. Adve. Foundations of the C++ concurrency memory model. In *PLDI '08: Proc. 2008 ACM SIGPLAN Conference on Programming Language Design and Implementation*, pages 68–78, June 2008. DOI: 10.1145/1375581.1375591 36, 65, 103

[44] Nathan G. Bronson, Christos Kozyrakis, and Kunle Olukotun. Feedback-directed barrier optimization in a strongly isolated STM. In *POPL '09: Proc. 36th Annual ACM SIGPLAN-SIGACT Symposium on Principles of Programming Languages*, pages 213–225, January 2009. DOI: 10.1145/1480881.1480909 71

[45] João Cachopo and António Rito-Silva. Versioned boxes as the basis for memory transactions. In *SCOOL '05: Proc. OOPSLA Workshop on Synchronization and Concurrency in Object-Oriented Languages*, October 2005. DOI: 10.1016/j.scico.2006.05.009 14, 143

[46] João Cachopo and António Rito-Silva. Combining software transactional memory with a domain modeling language to simplify web application development. In *ICWE '06: Proc. 6th International Conference on Web Engineering*, pages 297–304, July 2006. DOI: 10.1145/1145581.1145640 143

[47] Chi Cao Minh, JaeWoong Chung, Christos Kozyrakis, and Kunle Olukotun. STAMP: Stanford transactional applications for multi-processing. In *IISWC '08: Proceedings of The IEEE International Symposium on Workload Characterization*, September 2008. DOI: 10.1109/IISWC.2008.4636089 91

[48] Chi Cao Minh, Martin Trautmann, JaeWoong Chung, Austen McDonald, Nathan Bronson, Jared Casper, Christos Kozyrakis, and Kunle Olukotun. An effective hybrid transactional memory system with strong isolation guarantees. In *ISCA '07: Proc. 34th Annual International Symposium on Computer Architecture*, pages 69–80, June 2007. DOI: 10.1145/1250662.1250673 199

[49] Brian D. Carlstrom, JaeWoong Chung, Hassan Chafi, Austen McDonald, Chi Cao Minh, Lance Hammond, Christos Kozyrakis, and Kunle Olukotun. Transactional execution of Java programs. In *SCOOL '05: Proc. OOPSLA Workshop on Synchronization and Concurrency in Object-Oriented Languages*, October 2005. 86

[50] Brian D. Carlstrom, Austen McDonald, Hassan Chafi, JaeWoong Chung, Chi Cao Minh, Christos Kozyrakis, and Kunle Olukotun. The Atomos transactional programming language. In *PLDI '06: Proc. 2006 ACM SIGPLAN Conference on Programming Language Design and Implementation*, pages 1–13, June 2006. DOI: 10.1145/1133981.1133983 62, 74, 75, 81, 88

[51] Nuno Carvalho, João Cachopo, Luís Rodrigues, and António Rito-Silva. Versioned transactional shared memory for the FénixEDU web application. In *SDDDM '08: Proc. 2nd Workshop on Dependable Distributed Data Management*, pages 15–18, March 2008. DOI: 10.1145/1435523.1435526 143

[52] Calin Cascaval, Colin Blundell, Maged Michael, Harold W. Cain, Peng Wu, Stefanie Chiras, and Siddhartha Chatterjee. Software transactional memory: why is it only a research toy? *Communications of the ACM*, 51(11):40–46, November 2008. DOI: 10.1145/1400214.1400228 10

[53] Luis Ceze, James Tuck, Calin Cascaval, and Josep Torrellas. Bulk disambiguation of speculative threads in multiprocessors. In *ISCA '06: Proc. 33rd Annual International Symposium on Computer Architecture*, pages 227–238, June 2006. DOI: 10.1109/ISCA.2006.13 147, 170, 174, 176, 177

[54] Hassan Chafi, Chi Cao Minh, Austen McDonald, Brian D. Carlstrom, JaeWoong Chung, Lance Hammond, Christos Kozyrakis, and Kunle Olukotun. TAPE: A transactional application profiling environment. In *ICS '05: Proc. 19th Annual International Conference on Supercomputing*, pages 199–208, June 2005. DOI: 10.1145/1088149.1088176 90

[55] Hassan Chafi, Jared Casper, Brian D. Carlstrom, Austen McDonald, Chi Cao Minh, Woongki Baek, Christos Kozyrakis, and Kunle Olukotun. A scalable, non-blocking approach to transactional memory. In *HPCA '07: Proc. 13th International Symposium on High-Performance Computer Architecture*, pages 97–108, February 2007. DOI: 10.1109/HPCA.2007.346189 181

[56] David Chaiken, John Kubiatowicz, and Anant Agarwal. LimitLESS directories: a scalable cache coherence scheme. In *ASPLOS '91: Proc. 4th International Conference on Architectural Support for Programming Languages and Operating Systems*, pages 224–234, April 1991. DOI: 10.1145/106972.106995 157

[57] Albert Chang and Mark F. Mergen.   801 storage: architecture and programming. In *TOCS: ACM Transactions on Computer Systems*, volume 6, pages 28–50, 1988. DOI: 10.1145/35037.42270 194

[58] Shailender Chaudhry, Robert Cypher, Magnus Ekman, Martin Karlsson, Anders Landin, Sherman Yip, Håkan Zeffer, and Marc Tremblay. Rock: A high-performance Sparc CMT processor. *IEEE Micro*, 29(2):6–16, 2009. DOI: 10.1109/MM.2009.34 161

[59] Sigmund Cherem, Trishul Chilimbi, and Sumit Gulwani. Inferring locks for atomic sections. In *PLDI '08: Proc. 2008 ACM SIGPLAN Conference on Programming Language Design and Implementation*, pages 304–315, June 2008. DOI: 10.1145/1375581.1375619 9, 65

[60] Colin E. Cherry. Some experiments on the recognition of speech, with one and with two ears. *Journal of the Acoustical Society of America*, 25(5):975–979, 1953. DOI: 10.1121/1.1907229 2

[61] Dave Christie, Jae-Woong Chung, Stephan Diestelhorst, Michael Hohmuth, Martin Pohlack, Christof Fetzer, Martin Nowack, Torvald Riegel, Pascal Felber, Patrick Marlier, and Etienne Riviere. Evaluation of AMD's advanced synchronization facility within a complete transactional memory stack. In *EuroSys '10: Proc. 5th ACM European Conference on Computer Systems*, April 2010. DOI: 10.1145/1755913.1755918 13, 14, 149, 150, 158

[62] Weihaw Chuang, Satish Narayanasamy, Ganesh Venkatesh, Jack Sampson, Michael Van Biesbrouck, Gilles Pokam, Brad Calder, and Osvaldo Colavin. Unbounded page-based transactional memory. In *ASPLOS '06: Proc. 12th International Conference on Architectural Support for Programming Languages and Operating Systems*, pages 347–358, October 2006. DOI: 10.1145/1168857.1168901 196

[63] JaeWoong Chung, Chi Cao Minh, Brian D. Carlstrom, and Christos Kozyrakis. Parallelizing SPECjbb2000 with transactional memory. In *Proc. Workshop on Transactional Workloads*, June 2006. 57

[64] JaeWoong Chung, Chi Cao Minh, Austen McDonald, Travis Skare, Hassan Chafi, Brian D. Carlstrom, Christos Kozyrakis, and Kunle Olukotun.   Tradeoffs in transactional memory virtualization.   In *ASPLOS '06: Proc. 12th International Conference on Architectural Support for Programming Languages and Operating Systems*, pages 371–381, October 2006. DOI: 10.1145/1168857.1168903 196

[65] JaeWoong Chung, Hassan Chafi, Chi Cao Minh, Austen McDonald, Brian D. Carlstrom, Christos Kozyrakis, and Kunle Olukotun. The common case transactional behavior of multithreaded programs. In *HPCA '06: Proc. 12th International Symposium on High-Performance Computer Architecture*, pages 266–277, February 2006. DOI: 10.1109/HPCA.2006.1598135 90

[66] Cliff Click. HTM will not save the world, May 2010. Presentation at TMW10 workshop, `http://sss.cs.purdue.edu/projects/tm/tmw2010/Schedule.html`. 149, 150, 162

[67] E. G. Coffman, M. Elphick, and A. Shoshani. System deadlocks. *ACM Computing Surveys*, 3(2):67–78, 1971. DOI: 10.1145/356586.356588 158

[68] Ariel Cohen, John W. O'Leary, Amir Pnueli, Mark R. Tuttle, and Lenore D. Zuck. Verifying correctness of transactional memories. In *FMCAD '07: Proc. 7th International Conference on Formal Methods in Computer-Aided Design*, pages 37–44, November 2007. DOI: 10.1109/FAMCAD.2007.40 144

[69] Ariel Cohen, Amir Pnueli, and Lenore Zuck. Verification of transactional memories that support non-transactional memory accesses. In *TRANSACT '08: 3rd Workshop on Transactional Computing*, February 2008. 144

[70] Compaq. *Alpha architecture handbook*. October 1998. Version 4. 35, 155

[71] M. Couceiro, P. Romano, N. Carvalho, and L. Rodrigues. D2STM: Dependable distributed software transactional memory. In *PRDC '09: Proc. 15th Pacific Rim International Symposium on Dependable Computing*, November 2009. DOI: 10.1109/PRDC.2009.55 143

[72] A. L. Cox, S. Dwarkadas, P. Keleher, H. Lu, R. Rajamony, and W. Zwaenepoel. Software versus hardware shared-memory implementation: a case study. In *ISCA '94: Proc. 21st Annual International Symposium on Computer Architecture*, pages 106–117, April 1994. DOI: 10.1145/192007.192021 142

[73] Lawrence Crowl, Yossi Lev, Victor Luchangco, Mark Moir, and Dan Nussbaum. Integrating transactional memory into C++. In *TRANSACT '07: 2nd Workshop on Transactional Computing*, August 2007. 62

[74] Dave Cunningham, Khilan Gudka, and Susan Eisenbach. Keep off the grass: locking the right path for atomicity. In *CC '08: Proc. International Conference on Compiler Construction*, pages 276–290, March 2008. DOI: 10.1007/978-3-540-78791-4_19 9, 65

[75] Luke Dalessandro, Virendra J. Marathe, Michael F. Spear, and Michael L. Scott. Capabilities and limitations of library-based software transactional memory in C++. In *TRANSACT '07: 2nd Workshop on Transactional Computing*, August 2007. 94

[76] Luke Dalessandro and Michael L. Scott. Strong isolation is a weak idea. In *TRANSACT '09: 4th Workshop on Transactional Computing*, February 2009. 38, 41

[77] Luke Dalessandro, Michael F. Spear, and Michael L. Scott. NOrec: Streamlining STM by abolishing ownership records. In *PPoPP '10: Proc. 15th ACM Symposium on Principles and Practice of Parallel Programming*, pages 67–78, January 2010. DOI: 10.1145/1693453.1693464 15, 55, 102, 124, 126, 167

[78] Peter Damron, Alexandra Fedorova, Yossi Lev, Victor Luchangco, Mark Moir, and Dan Nussbaum. Hybrid transactional memory. In *ASPLOS '06: Proc. 12th International Conference on Architectural Support for Programming Languages and Operating Systems*, pages 336–346, October 2006. DOI: 10.1145/1168857.1168900 15, 165

[79] Alokika Dash and Brian Demsky. Software transactional distributed shared memory (poster). In *PPoPP '09: Proc. 14th ACM SIGPLAN Symposium on Principles and Practice of Parallel Programming*, pages 297–298, February 2009. DOI: 10.1145/1504176.1504223 144

[80] James C. Dehnert, Brian K. Grant, John P. Banning, Richard Johnson, Thomas Kistler, Alexander Klaiber, and Jim Mattson. The Transmeta code morphing software: using speculation, recovery, and adaptive retranslation to address real-life challenges. In *CGO '03: Proc. International Symposium on Code Generation and Optimization*, pages 15–24, March 2003. DOI: 10.1109/CGO.2003.1191529 161

[81] David Detlefs and Lingli Zhang. Transacting pointer-based accesses in an object-based software transactional memory system. In *TRANSACT '09: 4th Workshop on Transactional Computing*, February 2009. 68, 83, 105

[82] Dave Dice, Yossi Lev, Mark Moir, and Daniel Nussbaum. Early experience with a commercial hardware transactional memory implementation. In *ASPLOS '09: Proc. 14th International Conference on Architectural Support for Programming Languages and Operating Systems*, pages 157–168, March 2009. DOI: 10.1145/1508244.1508263 14, 149, 150, 161, 166

[83] Dave Dice, Ori Shalev, and Nir Shavit. Transactional locking II. In *DISC '06: Proc. 20th International Symposium on Distributed Computing*, pages 194–208, September 2006. Springer-Verlag Lecture Notes in Computer Science volume 4167. DOI: 10.1007/11864219_14 13, 15, 29, 47, 102, 105, 116, 122

[84] Dave Dice and Nir Shavit. Understanding tradeoffs in software transactional memory. In *CGO '07: Proc. International Symposium on Code Generation and Optimization*, pages 21–33, March 2007. DOI: 10.1109/CGO.2007.38 116

[85] David Dice and Nir Shavit. What really makes transactions faster? In *TRANSACT '06: 1st Workshop on Languages, Compilers, and Hardware Support for Transactional Computing*, June 2006. 47, 108, 116

[86] David Dice and Nir Shavit. TLRW: return of the read-write lock. In *TRANSACT '09: 4th Workshop on Transactional Computing*, February 2009. 10, 137

[87] Stephan Diestelhorst, Martin Pohlack, Michael Hohmuth, Dave Christie, Jae-Woong Chung, and Luke Yen. Implementing AMD's Advanced Synchronization Facility in an out-of-order x86 core. In *TRANSACT '10: 5th Workshop on Transactional Computing*, April 2010. 169

[88] Chen Ding, Xipeng Shen, Kirk Kelsey, Chris Tice, Ruke Huang, and Chengliang Zhang. Software behavior oriented parallelization. In *PLDI '07: Proc. 2007 ACM SIGPLAN Conference on Programming Language Design and Implementation*, pages 223–234, June 2007. DOI: 10.1145/1250734.1250760 xiii, 55, 99, 128

[89] Simon Doherty, Lindsay Groves, Victor Luchangco, and Mark Moir. Towards formally specifying and verifying transactional memory. In *Refinement Workshop 2009*, November 2009. 23, 145

[90] Shlomi Dolev, Danny Hendler, and Adi Suissa. CAR-STM: scheduling-based collision avoidance and resolution for software transactional memory. In *PODC '08: Proc. 27th ACM Symposium on Principles of Distributed Computing*, pages 125–134, August 2008. DOI: 10.1145/1400751.1400769 53

[91] Kevin Donnelly and Matthew Fluet. Transactional events. In *ICFP '06: Proc. 11th ACM SIGPLAN International Conference on Functional Programming*, pages 124–135, September 2006. DOI: 10.1145/1159803.1159821 92

[92] Aleksandar Dragojević, Pascal Felber, Vincent Gramoli, and Rachid Guerraoui. Why STM can be more than a research toy. Technical Report LPD-REPORT-2009-003, LPD (Distributed Programming Laboratory), EPFL, 2009. 10

[93] Aleksandar Dragojević, Rachid Guerraoui, and Michał Kapałka. Stretching transactional memory. In *PLDI '09: Proc. 2009 ACM SIGPLAN Conference on Programming Language Design and Implementation*, pages 155–165, June 2009. DOI: 10.1145/1542476.1542494 15, 52, 123

[94] Aleksandar Dragojević, Rachid Guerraoui, Anmol V. Singh, and Vasu Singh. Preventing versus curing: avoiding conflicts in transactional memories. In *PODC '09: Proc. 28th ACM Symposium on Principles of Distributed Computing*, pages 7–16, August 2009. DOI: 10.1145/1582716.1582725 54

[95] Polina Dudnik and Michael Swift. Condition variables and transactional memory: problem or opportunity? In *TRANSACT '09: 4th Workshop on Transactional Computing*, February 2009. 86

[96] Laura Effinger-Dean, Matthew Kehrt, and Dan Grossman. Transactional events for ML. In *ICFP '08: Proc. 13th ACM SIGPLAN International Conference on Functional Programming*, pages 103–114, September 2008. DOI: 10.1145/1411204.1411222 92

[97] Faith Ellen, Yossi Lev, Victor Luchangco, and Mark Moir. SNZI: scalable nonzero indicators. In *PODC '07: Proc. 26th ACM Symposium on Principles of Distributed Computing*, pages 13–22, August 2007. DOI: 10.1145/1281100.1281106 115, 140

[98] Tayfun Elmas, Shaz Qadeer, and Serdar Tasiran. Goldilocks: a race and transaction-aware Java runtime. In *PLDI '07: Proc. 2007 ACM SIGPLAN Conference on Programming Language Design and Implementation*, pages 245–255, June 2007. DOI: 10.1145/1250734.1250762 90

[99] Robert Ennals. Efficient software transactional memory. Technical Report IRC-TR-05-051, Intel Research Cambridge Tech Report, January 2005. 47, 116

[100] Robert Ennals. Software transactional memory should not be obstruction-free. Technical Report IRC-TR-06-052, Intel Research Cambridge Tech Report, January 2006. 47

[101] K. P. Eswaran, J. N. Gray, R. A. Lorie, and I. L. Traiger. The notions of consistency and predicate locks in a database system. *Communications of the ACM*, 19(11):624–633, 1976. DOI: 10.1145/360363.360369 62

[102] Li Fan, Pei Cao, Jussara M. Almeida, and Andrei Z. Broder. Summary cache: a scalable wide-area web cache sharing protocol. *IEEE/ACM Transactions on Networking*, 8(3):281–293, 2000. DOI: 10.1109/90.851975 193

[103] Pascal Felber, Christof Fetzer, and Torvald Riegel. Dynamic performance tuning of word-based software transactional memory. In *PPoPP '08: Proc. 13th ACM SIGPLAN Symposium on Principles and Practice of Parallel Programming*, pages 237–246, February 2008. DOI: 10.1145/1345206.1345241 107, 119, 120

[104] Pascal Felber, Vincent Gramoli, and Rachid Guerraoui. Elastic transactions. In *DISC '09: Proc. 23rd International Symposium on Distributed Computing*, September 2009. DOI: 10.1007/978-3-642-04355-0_12 56

[105] Keir Fraser. *Practical lock freedom*. PhD thesis, Cambridge University Computer Laboratory, 2003. Also available as Technical Report UCAM-CL-TR-579. 55, 56, 70, 82, 131, 137

[106] Keir Fraser and Tim Harris. Concurrent programming without locks. *TOCS: ACM Transactions on Computer Systems*, 25(2), May 2007. DOI: 10.1145/1233307.1233309 17, 131

[107] Vladimir Gajinov, Ferad Zyulkyarov, Osman S. Unsal, Adrián Cristal, Eduard Ayguadé, Tim Harris, and Mateo Valero. QuakeTM: parallelizing a complex sequential application using transactional memory. In *ICS '09: Proc. 23rd International Conference on Supercomputing*, pages 126–135, June 2009. DOI: 10.1145/1542275.1542298 43, 92

[108] Justin Gottschlich and Daniel A. Connors. DracoSTM: a practical C++ approach to software transactional memory. In *LCSD '07: Proc. 2007 ACM SIGPLAN Symposium on Library-Centric Software Design*, October 2007. DOI: 10.1145/1512762.1512768 94

[109] Justin Gottschlich and Daniel A. Connors. Extending contention managers for user-defined priority-based transactions. In *EPHAM '08: Workshop on Exploiting Parallelism with Transactional Memory and other Hardware Assisted Methods*, April 2008. 53

[110] Justin E. Gottschlich, Jeremy G. Siek, and Daniel A. Connors. C++ move semantics for exception safety and optimization in software transactional memory libraries. In *ICOOOLPS '08: Proc. 3rd International Workshop on Implementation, Compilation, Optimization of Object-Oriented Languages, Programs and Systems*. July 2008. 94

[111] Justin E. Gottschlich, Jeremy G. Siek, Manish Vachharajani, Dwight Y. Winkler, and Daniel A. Connors. An efficient lock-aware transactional memory implementation. In *ICOOOLPS '09: Proc. 4th workshop on the Implementation, Compilation, Optimization of Object-Oriented Languages and Programming Systems*, pages 10–17, July 2009. DOI: 10.1145/1565824.1565826 85

[112] Justin E. Gottschlich, Manish Vachharajani, and Siek G. Jeremy. An efficient software transactional memory using commit-time invalidation. In *CGO '10: Proc. International Symposium on Code Generation and Optimization*, apr 2010. DOI: 10.1145/1772954.1772970 125

[113] Jim Gray and Andreas Reuter. *Transaction Processing: Concepts and Techniques*. Morgan Kaufmann, 1993. 5

[114] Dan Grossman. The transactional memory / garbage collection analogy. In *OOPSLA '07: Proc. 22nd ACM SIGPLAN Conference on Object-Oriented Programming, Systems, Languages, and Applications (Essays)*, pages 695–706, October 2007. DOI: 10.1145/1297027.1297080 64

[115] Dan Grossman, Jeremy Manson, and William Pugh. What do high-level memory models mean for transactions? In *MSPC '06: Proc. 2006 Workshop on Memory System Performance and Correctness*, October 2006. DOI: 10.1145/1178597.1178609 23, 31, 38, 136

[116] A. S. Grove. *Only the paranoid survive*. Doubleday, 1996. 1

[117] Rachid Guerraoui. *Transactional Memory: The Theory*. Morgan & Claypool. Synthesis Lectures on Distributed Computing Theory (to appear). xiii

[118] Rachid Guerraoui, Thomas Henzinger, and Vasu Singh. Completeness and nondeterminism in model checking transactional memories. In *CONCUR '08: Proc. 19th Confernece on Concurrency Theory*, pages 21–35, August 2008. DOI: 10.1007/978-3-540-85361-9_6 144

[119] Rachid Guerraoui, Thomas Henzinger, and Vasu Singh. Permissiveness in transactional memories. In *DISC '08: Proc. 22nd International Symposium on Distributed Computing*, pages 305–319, September 2008. Springer-Verlag Lecture Notes in Computer Science volume 5218. DOI: 10.1007/978-3-540-87779-0_21 23, 48

[120] Rachid Guerraoui, Thomas A. Henzinger, Barbara Jobstmann, and Vasu Singh. Model checking transactional memories. In *PLDI '08: Proc. 2008 ACM SIGPLAN Conference on Programming Language Design and Implementation*, pages 372–382, June 2008. DOI: 10.1145/1375581.1375626 144

[121] Rachid Guerraoui, Thomas A. Henzinger, and Vasu Singh. Software transactional memory on relaxed memory models. In *CAV '09: Proc. 21st International Conference on Computer Aided Verification*, 2009. DOI: 10.1007/978-3-642-02658-4_26 103

[122] Rachid Guerraoui, Maurice Herlihy, and Bastian Pochon. Polymorphic contention management. In *DISC '05: Proc. 19th International Symposium on Distributed Computing*, pages 303–323. LNCS, Springer, September 2005. DOI: 10.1007/11561927_23 51, 131

[123] Rachid Guerraoui, Maurice Herlihy, and Bastian Pochon. Toward a theory of transactional contention managers. In *PODC '05: Proc. 24th Annual ACM SIGACT-SIGOPS Symposium on Principles of Distributed Computing*, pages 258–264, July 2005. DOI: 10.1145/1073814.1073863 48, 49, 51

[124] Rachid Guerraoui and Michał Kapałka. On obstruction-free transactions. In *SPAA '08: Proc. 20th Annual Symposium on Parallelism in Algorithms and Architectures*, pages 304–313, June 2008. DOI: 10.1145/1378533.1378587 48

[125] Rachid Guerraoui and Michał Kapałka. On the correctness of transactional memory. In *PPoPP '08: Proc. 13th ACM SIGPLAN Symposium on Principles and Practice of Parallel Programming*, pages 175–184, February 2008. DOI: 10.1145/1345206.1345233 23, 29

[126] Rachid Guerraoui and Michał Kapałka. The semantics of progress in lock-based transactional memory. In *POPL '09: Proc. 36th Annual ACM SIGPLAN-SIGACT Symposium on Principles of Programming Languages*, pages 404–415, January 2009. DOI: 10.1145/1480881.1480931 23

[127] Rachid Guerraoui, Michał Kapałka, and Jan Vitek. STMBench7: A benchmark for software transactional memory. In *EuroSys '07: Proc. 2nd ACM European Conference on Computer Systems*, pages 315–324, March 2007. DOI: 10.1145/1272998.1273029 90

[128] Nicholas Haines, Darrell Kindred, J. Gregory Morrisett, Scott M. Nettles, and Jeannette M. Wing. Composing first-class transactions. *TOPLAS: ACM Transactions on Programming Languages and Systems*, 16(6):1719–1736, 1994. DOI: 10.1145/197320.197346 42

[129] Richard L. Halpert, Christopher J. F. Pickett, and Clark Verbrugge. Component-based lock allocation. In *PACT '07: Proc. 16th International Conference on Parallel Architecture and Compilation Techniques*, pages 353–364, September 2007. DOI: 10.1109/PACT.2007.23 9, 65

[130] Robert H. Halstead, Jr. MULTILISP: a language for concurrent symbolic computation. *TOPLAS: ACM Transactions on Programming Languages and Systems*, 7(4):501–538, 1985. DOI: 10.1145/4472.4478 98

[131] Lance Hammond, Brian D. Carlstrom, Vicky Wong, Ben Hertzberg, Mike Chen, Christos Kozyrakis, and Kunle Olukotun. Programming with transactional coherence and consistency (TCC). In *ASPLOS '04: Proc. 11th International Conference on Architectural Support for Programming Languages and Operating Systems*, pages 1–13, October 2004. DOI: 10.1145/1024393.1024395 94

[132] Lance Hammond, Vicky Wong, Mike Chen, Brian D. Carlstrom, John D. Davis, Ben Hertzberg, Manohar K. Prabhu, Honggo Wijaya, Christos Kozyrakis, and Kunle Olukotun. Transactional memory coherence and consistency. In *ISCA '04: Proc. 31st Annual International Symposium on Computer Architecture*, page 102, June 2004. DOI: 10.1145/1028176.1006711 147, 170, 179

[133] Tim Harris. Exceptions and side-effects in atomic blocks. In *CSJP '04: Proc. ACM PODC Workshop on Concurrency and Synchronization in Java Programs*, pages 46–53, July 2004. Proceedings published as Memorial University of Newfoundland CS Technical Report 2004-01. DOI: 10.1016/j.scico.2005.03.005 81, 88

[134] Tim Harris and Keir Fraser. Language support for lightweight transactions. In *OOPSLA '03: Proc. Object-Oriented Programming, Systems, Languages, and Applications*, pages 388–402, October 2003. DOI: 10.1145/949343.949340 62, 68, 75, 83, 133, 141

[135] Tim Harris and Keir Fraser. Revocable locks for non-blocking programming. In *Proc. ACM Symposium on Principles and Practice of Parallel Programming*, June 2005. DOI: 10.1145/1065944.1065954 55, 128, 134

[136] Tim Harris, Maurice Herlihy, Simon Marlow, and Simon Peyton Jones. Composable memory transactions. In *PPoPP '05: Proc. ACM Symposium on Principles and Practice of Parallel Programming*, June 2005. A shorter version appeared in *CACM* 51(8):91–100, August 2008. DOI: 10.1145/1065944.1065952 40, 62, 68, 73, 74, 83, 84

[137] Tim Harris and Simon Peyton Jones. Transactional memory with data invariants. In *TRANSACT '06: 1st Workshop on Languages, Compilers, and Hardware Support for Transactional Computing*, June 2006. 24

[138] Tim Harris, Mark Plesko, Avraham Shinnar, and David Tarditi. Optimizing memory transactions. In *PLDI '06: Proc. 2006 ACM SIGPLAN Conference on Programming Language Design and Implementation*, pages 14–25, June 2006. DOI: 10.1145/1133981.1133984 47, 62, 68, 83, 102, 105, 106, 107, 109, 113, 114

[139] Tim Harris and Srdjan Stipic. Abstract nested transactions. In *TRANSACT '07: 2nd Workshop on Transactional Computing*, August 2007. 55, 57

[140] Tim Harris, Sasa Tomic, Adrián Cristal, and Osman Unsal. Dynamic filtering: multi-purpose architecture support for language runtime systems. In *ASPLOS '10: Proc. 15th International*

Conference on Architectural Support for Programming Language and Operating Systems, pages 39–52, March 2010. DOI: 10.1145/1736020.1736027 198

[141] Maurice Herlihy. Wait-free synchronization. *TOPLAS: ACM Transactions on Programming Languages and Systems*, 13(1):124–149, January 1991. DOI: 10.1145/114005.102808 47

[142] Maurice Herlihy and Eric Koskinen. Transactional boosting: a methodology for highly-concurrent transactional objects. In *PPoPP '08: Proc. 13th ACM SIGPLAN Symposium on Principles and Practice of Parallel Programming*, pages 207–216, February 2008. DOI: 10.1145/1345206.1345237 58

[143] Maurice Herlihy and Yossi Lev. tm_db: a generic debugging library for transactional programs. In *PACT '09: Proc. 18th International Conference on Parallel Architectures and Compilation Techniques*, pages 136–145, September 2009. DOI: 10.1109/PACT.2009.23 55, 89

[144] Maurice Herlihy, Victor Luchangco, and Mark Moir. The repeat offender problem: a mechanism for supporting dynamic-sized, lock-free data structures. In *DISC '02: Proceedings of the 16th International Conference on Distributed Computing*, pages 339–353, 2002. DOI: 10.1007/3-540-36108-1_23 82

[145] Maurice Herlihy, Victor Luchangco, and Mark Moir. Obstruction-free synchronization: double-ended queues as an example. In *ICDCS '03: Proc. 23rd International Conference on Distributed Computing Systems*, pages 522–529, May 2003. DOI: 10.1109/ICDCS.2003.1203503 47

[146] Maurice Herlihy, Victor Luchangco, Mark Moir, and William N. Scherer III. Software transactional memory for dynamic-sized data structures. In *PODC '03: Proc. 22nd ACM Symposium on Principles of Distributed Computing*, pages 92–101, July 2003. DOI: 10.1145/872035.872048 56, 94, 128, 131

[147] Maurice Herlihy, Mark Moir, and Victor Luchangco. A flexible framework for implementing software transactional memory. In *OOPSLA '06: Proc. 21st ACM SIGPLAN Conference on Object-Oriented Programing, Systems, Languages, and Applications*, pages 253–262, October 2006. DOI: 10.1145/1167473.1167495 131

[148] Maurice Herlihy and J. Eliot B. Moss. Transactional memory: architectural support for lock-free data structures. In *ISCA '93: Proc. 20th Annual International Symposium on Computer Architecture*, pages 289–300, May 1993. DOI: 10.1145/165123.165164 6, 17, 149, 155

[149] Maurice Herlihy and Nir Shavit. *The art of multiprocessor programming*. Morgan Kaufmann, 2008. DOI: 10.1145/1146381.1146382 56

[150] Maurice Herlihy and Jeannette M. Wing. Linearizability: a correctness condition for concurrent objects. *TOPLAS: ACM Transactions on Programming Languages and Systems*, 12(3):463–492, July 1990. DOI: 10.1145/78969.78972 23

[151] Michael Hicks, Jeffrey S. Foster, and Polyvios Prattikakis. Lock inference for atomic sections. In *TRANSACT '06: 1st Workshop on Languages, Compilers, and Hardware Support for Transactional Computing*, June 2006. 9, 65

[152] Mark D. Hill, Derek Hower, Kevin E. Moore, Michael M. Swift, Haris Volos, and David A. Wood. A case for deconstructing hardware transactional memory systems. Technical Report CS-TR-2007-1594, University of Wisconsin-Madison, 2007. Also Dagstuhl Seminar Proceedings 07361. 169

[153] W. Daniel Hillis and Guy L. Steele,Jr. Data parallel algorithms. *Communications of the ACM*, 29(12):1170–1183, 1986. DOI: 10.1145/7902.7903 2

[154] Benjamin Hindman and Dan Grossman. Atomicity via source-to-source translation. In *MSPC '06: Proc. 2006 Workshop on Memory System Performance and Correctness*, October 2006. DOI: 10.1145/1178597.1178611 71, 116

[155] Benjamin Hindman and Dan Grossman. Strong atomicity for Java without virtual-machine support. Technical Report 2006-05-01, University of Washington, Dept. Computer Science, May 2006. 35, 71

[156] C. A. R. Hoare. Monitors: an operating system structuring concept. *Communications of the ACM*, pages 549–557, October 1974. DOI: 10.1145/355620.361161 63

[157] C. A. R. Hoare. Towards a theory of parallel programming. In *The origin of concurrent programming: from semaphores to remote procedure calls*, pages 231–244. Springer-Verlag, 2002. DOI: 10.1007/3-540-07994-7_47 63, 76

[158] Owen S. Hofmann, Donald E. Porter, Christopher J. Rossbach, Hany E. Ramadan, and Emmett Witchel. Solving difficult HTM problems without difficult hardware. In *TRANSACT '07: 2nd Workshop on Transactional Computing*, August 2007. 203

[159] Owen S. Hofmann, Christopher J. Rossbach, and Emmett Witchel. Maximum benefit from a minimal HTM. In *ASPLOS '09: Proc. 14th International Conference on Architectural Support for Programming Languages and Operating Systems*, pages 145–156, March 2009. DOI: 10.1145/1508244.1508262 203

[160] James J. Horning, Hugh C. Lauer, P. M. Melliar-Smith, and Brian Randell. A program structure for error detection and recovery. In *Operating Systems, Proceedings of an International Symposium*, pages 171–187. Springer-Verlag, 1974. DOI: 10.1007/BFb0029359 80

[161] Liyang Hu and Graham Hutton. Towards a verified implementation of software transactional memory. In *Proc. Symposium on Trends in Functional Programming*, May 2008. 145

[162] Richard L. Hudson, Bratin Saha, Ali-Reza Adl-Tabatabai, and Benjamin C. Hertzberg. McRT-Malloc: a scalable transactional memory allocator. In *ISMM '06: Proc. 5th International Symposium on Memory Management*, pages 74–83, June 2006. DOI: 10.1145/1133956.1133967 82, 105, 137

[163] Damien Imbs, José Ramon de Mendivil, and Michel Raynal. Virtual world consistency: a new condition for STM systems (brief announcement). In *PODC '09: Proc. 28th ACM Symposium on Principles of Distributed Computing*, pages 280–281, August 2009. DOI: 10.1145/1582716.1582764 30

[164] Amos Israeli and Lihu Rappoport. Disjoint-access-parallel implementations of strong shared memory primitives. In *PODC '94: Proc. 13th ACM Symposium on Principles of Distributed Computing*, pages 151–160, August 1994. DOI: 10.1145/197917.198079 48

[165] Eric H. Jensen, Gary W. Hagensen, and Jeffrey M. Broughton. A new approach to exclusive data access in shared memory multiprocessors. Technical Report Technical Report UCRL-97663, November 1987. 155

[166] Gerry Kane. *MIPS RISC architecture*. Prentice-Hall, Inc., Upper Saddle River, NJ, USA, 1988. 155

[167] Seunghwa Kang and David A. Bader. An efficient transactional memory algorithm for computing minimum spanning forest of sparse graphs. In *PPoPP '09: Proc. 14th ACM SIGPLAN Symposium on Principles and Practice of Parallel Programming*, pages 15–24, February 2009. DOI: 10.1145/1504176.1504182 91

[168] Gokcen Kestor, Srdjan Stipic, Osman S. Unsal, Adrián Cristal, and Mateo Valero. RMS-TM: A transactional memory benchmark for recognition, mining and synthesis applications. In *TRANSACT '09: 4th Workshop on Transactional Computing*, February 2009. 91

[169] Behram Khan, Matthew Horsnell, Ian Rogers, Mikel Luján, Andrew Dinn, and Ian Watson. An object-aware hardware transactional memory. In *HPCC '08: Proc. 10th International Conference on High Performance Computing and Communications*, pages 93–102, September 2008. DOI: 10.1109/HPCC.2008.110 150

[170] Aaron Kimball and Dan Grossman. Software transactions meet first-class continuations. In *Proc. 8th Annual Workshop on Scheme and Functional Programming*, September 2007. 62, 80

[171] Thomas F. Knight. An architecture for mostly functional languages. In *LFP '86: Proc. ACM Lisp and Functional Programming Conference*, pages 500–519, August 1986. DOI: 10.1145/319838.319854 179

[172] G.K. Konstadinidis, M. Tremblay, S. Chaudhry, M. Rashid, P.F. Lai, Y. Otaguro, Y. Orginos, S. Parampalli, M. Steigerwald, S. Gundala, R. Pyapali, L.D. Rarick, I. Elkin, Y. Ge, and

I. Parulkar. Architecture and physical implementation of a third generation 65nm, 16 core, 32 thread chip-multithreading SPARC processor. *IEEE Journal of Solid-State Circuits*, 44(1):7–17, January 2009. DOI: 10.1109/JSSC.2008.2007144 161

[173] Guy Korland, Nir Shavit, and Pascal Felber. Noninvasive Java concurrency with Deuce STM (poster). In *SYSTOR '09: The Israeli Experimental Systems Conference*, May 2009. Further details at http://www.deucestm.org/. 13

[174] Guy Korland, Nir Shavit, and Pascal Felber. Noninvasive concurrency with Java STM. In *MULTIPROG '10: Proc. 3rd Workshop on Programmability Issues for Multi-Core Computers*, January 2010. 68, 131

[175] Eric Koskinen and Maurice Herlihy. Checkpoints and continuations instead of nested transactions. In *SPAA '08: Proc. 20th Annual Symposium on Parallelism in Algorithms and Architectures*, pages 160–168, June 2008. An earlier version appeared at *TRANSACT '08*. DOI: 10.1145/1378533.1378563 57

[176] Eric Koskinen and Maurice Herlihy. Dreadlocks: efficient deadlock detection. In *SPAA '08: Proc. 20th Annual Symposium on Parallelism in Algorithms and Architectures*, pages 297–303, June 2008. An earlier version appeared at *TRANSACT '08*. DOI: 10.1145/1378533.1378585 21, 109

[177] Eric Koskinen and Maurice Herlihy. Concurrent non-commutative boosted transactions. In *TRANSACT '09: 4th Workshop on Transactional Computing*, February 2009. 58

[178] Eric Koskinen, Matthew Parkinson, and Maurice Herlihy. Coarse-grained transactions. In *POPL '10: Proc. 37th Annual ACM SIGPLAN-SIGACT Symposium on Principles of Programming Languages*, pages 19–30, January 2010. DOI: 10.1145/1706299.1706304 58

[179] Christos Kotselidis, Mohammad Ansari, Kim Jarvis, Mikel Luján, Chris C. Kirkham, and Ian Watson. DiSTM: A software transactional memory framework for clusters. In *ICPP '08: Proc. 37th International Conference on Parallel Processing*, September 2008. DOI: 10.1109/ICPP.2008.59 143

[180] Christos Kotselidis, Mohammad Ansari, Kim Jarvis, Mikel Luján, Chris C. Kirkham, and Ian Watson. Investigating software transactional memory on clusters. In *IPDPS '08: Proc. 22nd International Parallel and Distributed Processing Symposium*, 2008. DOI: 10.1109/IPDPS.2008.4536340 143

[181] Milind Kulkarni, Patrick Carribault, Keshav Pingali, Ganesh Ramanarayanan, Bruce Walter, Kavita Bala, and L. Paul Chew. Scheduling strategies for optimistic parallel execution of irregular programs. In *SPAA '08: Proc. 20th Annual Symposium on Parallelism in Algorithms and Architectures*, pages 217–228, June 2008. DOI: 10.1145/1378533.1378575 54, 58

[182] Milind Kulkarni, Keshav Pingali, Bruce Walter, Ganesh Ramanarayanan, Kavita Bala, and L. Paul Chew. Optimistic parallelism requires abstractions. In *PLDI '07: Proc. 2007 ACM SIGPLAN Conference on Programming Language Design and Implementation*, pages 211–222, June 2007. DOI: 10.1145/1250734.1250759 58

[183] Sanjeev Kumar, Michael Chu, Christopher J. Hughes, Partha Kundu, and Anthony Nguyen. Hybrid transactional memory. In *PPoPP '06: Proc. 11th ACM SIGPLAN Symposium on Principles and Practice of Parallel Programming*, March 2006. DOI: 10.1145/1122971.1123003 166

[184] Bradley C. Kuszmaul and Charles E. Leiserson. Transactions everywhere, January 2003. Available at http://dspace.mit.edu/handle/1721.1/3692. 94

[185] Edmund S. L. Lam and Martin Sulzmann. A concurrent constraint handling rules implementation in Haskell with software transactional memory. In *DAMP '07: Proc. 2007 workshop on Declarative aspects of multicore programming*, pages 19–24, January 2007. DOI: 10.1145/1248648.1248653 92

[186] Leslie Lamport. Time, clocks, and the ordering of events in a distributed system. *Communications of the ACM*, 21(7):558–565, 1978. DOI: 10.1145/359545.359563 161

[187] Leslie Lamport. How to make a multiprocessor computer that correctly executes multiprocess progranm. *IEEE Trans. Comput.*, 28(9):690–691, 1979. DOI: 10.1109/TC.1979.1675439 38

[188] Yossi Lev, Victor Luchangco, Virendra Marathe, Mark Moir, Dan Nussbaum, and Marek Olszewski. Anatomy of a scalable software transactional memory. In *TRANSACT '09: 4th Workshop on Transactional Computing*, February 2009. 15, 115, 122, 131, 139, 140

[189] Yossi Lev and Jan-Willem Maessen. Toward a safer interaction with transactional memory by tracking object visibility. In *SCOOL '05: Proc. OOPSLA Workshop on Synchronization and Concurrency in Object-Oriented Languages*, October 2005. 40

[190] Yossi Lev and Jan-Willem Maessen. Split hardware transactions: true nesting of transactions using best-effort hardware transactional memory. In *PPoPP '08: Proc. 13th ACM SIGPLAN Symposium on Principles and Practice of Parallel Programming*, pages 197–206, February 2008. DOI: 10.1145/1345206.1345236 81, 88, 202

[191] Yossi Lev and Mark Moir. Fast read sharing mechanism for software transactional memory (poster). In *PODC '04: Proc. 23rd ACM Symposium on Principles of Distributed Computing*, 2004. 115, 131

[192] Yossi Lev and Mark Moir. Debugging with transactional memory. In *TRANSACT '06: 1st Workshop on Languages, Compilers, and Hardware Support for Transactional Computing*, June 2006. 89

[193] Yossi Lev, Mark Moir, and Dan Nussbaum. PhTM: Phased transactional memory. In *TRANSACT '07: 2nd Workshop on Transactional Computing*, August 2007. 15, 166

[194] Sean Lie. Hardware support for unbounded transactional memory. Master's thesis, May 2004. Massachusetts Institute of Technology. 165

[195] Richard J. Lipton. Reduction: a method of proving properties of parallel programs. *Communications of the ACM*, 18(12):717–721, December 1975. DOI: 10.1145/361227.361234 58

[196] Barbara Liskov. Distributed programming in Argus. *Communications of the ACM*, 31(3):300–312, 1988. DOI: 10.1145/42392.42399 5

[197] Yi Liu, Xin Zhang, He Li, Mingxiu Li, and Depei Qian. Hardware transactional memory supporting I/O operations within transactions. In *HPCC '08: Proc. 10th International Conference on High Performance Computing and Communications*, pages 85–92, September 2008. DOI: 10.1109/HPCC.2008.71 81

[198] Yu David Liu, Xiaoqi Lu, and Scott F. Smith. Coqa: concurrent objects with quantized atomicity. In *CC '08: Proc. International Conference on Compiler Construction*, pages 260–275, March 2008. DOI: 10.1007/978-3-540-78791-4_18 96

[199] David B. Lomet. Process structuring, synchronization, and recovery using atomic actions. In *ACM Conference on Language Design for Reliable Software*, pages 128–137, March 1977. DOI: 10.1145/800022.808319 6, 62

[200] João Lourenço and Goncalo T. Cunha. Testing patterns for software transactional memory engines. In *PADTAD '07: Proc. 2007 ACM Workshop on Parallel and Distributed Systems: Testing and Debugging*, pages 36–42, July 2007. DOI: 10.1145/1273647.1273655 144

[201] David B. Loveman. High performance Fortran. *IEEE Parallel Distrib. Technol.*, 1(1):25–42, 1993. DOI: 10.1109/88.219857 2

[202] Victor Luchangco. Against lock-based semantics for transactional memory (brief announcement). In *SPAA '08: Proc. 20th Symposium on Parallelism in Algorithms and Architectures*, pages 98–100, June 2008. DOI: 10.1145/1378533.1378549 36

[203] Victor Luchangco and Virendra J. Marathe. Transaction synchronizers. In *SCOOL '05: Proc. OOPSLA Workshop on Synchronization and Concurrency in Object-Oriented Languages*, October 2005. 76

[204] Marc Lupon, Grigorios Magklis, and Antonio González. Version management alternatives for hardware transactional memory. In *MEDEA '08: Proc. 9th Workshop on Memory Performance*, October 2008. DOI: 10.1145/1509084.1509094 172

[205] Kaloian Manassiev, Madalin Mihailescu, and Cristiana Amza. Exploiting distributed version concurrency in a transactional memory cluster. In *PPoPP '06: Proc. 11th ACM SIGPLAN Symposium on Principles and Practice of Parallel Programming*, pages 198–208, March 2006. DOI: 10.1145/1122971.1123002 142

[206] Chaiyasit Manovit, Sudheendra Hangal, Hassan Chafi, Austen McDonald, Christos Kozyrakis, and Kunle Olukotun. Testing implementations of transactional memory. In *PACT '06: Proc. 15th international conference on Parallel architectures and compilation techniques*, pages 134–143, September 2006. DOI: 10.1145/1152154.1152177 144

[207] Jeremy Manson, William Pugh, and Sarita V. Adve. The Java memory model. In *POPL '05: Proc. 32nd ACM SIGPLAN-SIGACT Symposium on Principles of Programming Languages*, pages 378–391, January 2005. DOI: 10.1145/1040305.1040336 34, 35, 36, 65, 97, 103

[208] Virendra J. Marathe and Mark Moir. Toward high performance nonblocking software transactional memory. In *PPoPP '08: Proc. 13th ACM SIGPLAN Symposium on Principles and Practice of Parallel Programming*, pages 227–236, February 2008. DOI: 10.1145/1345206.1345240 47, 102, 134

[209] Virendra J. Marathe, William N. Scherer III, and Michael L. Scott. Adaptive software transactional memory. In *Proc. 19th International Symposium on Distributed Computing*, September 2005. Earlier but expanded version available as TR 868, University of Rochester Computer Science Dept., May 2005. 131, 132

[210] Virendra J. Marathe and Michael L. Scott. Using LL/SC to simplify word-based software transactional memory (poster). In *PODC '05: Proc. 24th ACM Symposium on Principles of Distributed Computing*, July 2005. 135

[211] Virendra J. Marathe, Michael F. Spear, Christopher Heriot, Athul Acharya, David Eisenstat, William N. Scherer III, and Michael L. Scott. Lowering the overhead of software transactional memory. Technical Report TR 893, Computer Science Department, University of Rochester, March 2006. Condensed version presented at TRANSACT '06. 15, 131, 132, 197, 198

[212] Virendra J. Marathe, Michael F. Spear, and Michael L. Scott. Scalable techniques for transparent privatization in software transactional memory. In *ICPP '08: Proc. 37th International Conference on Parallel Processing*, September 2008. DOI: 10.1109/ICPP.2008.69 138, 139

[213] Milo M. K. Martin, Mark D. Hill, and David A. Wood. Token coherence: decoupling performance and correctness. In *ISCA '03: Proc. 30th International Symposium on Computer Architecture*, pages 182–193, June 2003. DOI: 10.1109/ISCA.2003.1206999 190

[214] Bill McCloskey, Feng Zhou, David Gay, and Eric Brewer. Autolocker: synchronization inference for atomic sections. In *POPL '06: Proc. 33rd ACM SIGPLAN-*

SIGACT Symposium on Principles of Programming Languages, pages 346–358, January 2006. DOI: 10.1145/1111037.1111068 9, 65

[215] Austen McDonald, JaeWoong Chung, Brian D. Carlstrom, Chi Cao Minh, Hassan Chafi, Christos Kozyrakis, and Kunle Olukotun. Architectural semantics for practical transactional memory. In *ISCA '06: Proc. 33rd Annual International Symposium on Computer Architecture*, pages 53–65, June 2006. 81, 88, 201, 202

[216] Phil McGachey, Ali-Reza Adl-Tabatabai, Richard L. Hudson, Vijay Menon, Bratin Saha, and Tatiana Shpeisman. Concurrent GC leveraging transactional memory. In *PPoPP '08: Proc. 13th ACM SIGPLAN Symposium on Principles and Practice of Parallel Programming*, pages 217–226, February 2008. DOI: 10.1145/1345206.1345238 83

[217] P.E. McKenney and J. D. Slingwine. Read-copy update: using execution history to solve concurrency problems. In *Proc. 10th International Conference on Parallel and Distributed Computing and Systems*, pages 508–518, 1998. 82

[218] Mojtaba Mehrara, Jeff Hao, Po-Chun Hsu, and Scott Mahlke. Parallelizing sequential applications on commodity hardware using a low-cost software transactional memory. In *PLDI '09: Proc. 2009 ACM SIGPLAN Conference on Programming Language Design and Implementation*, pages 166–176, June 2009. DOI: 10.1145/1542476.1542495 125

[219] Vijay Menon, Steven Balensiefer, Tatiana Shpeisman, Ali-Reza Adl-Tabatabai, Richard Hudson, Bratin Saha, and Adam Welc. Single global lock semantics in a weakly atomic STM. In *TRANSACT '08: 3rd Workshop on Transactional Computing*, February 2008. 31, 37, 38, 65, 136, 137

[220] Vijay Menon, Steven Balensiefer, Tatiana Shpeisman, Ali-Reza Adl-Tabatabai, Richard L. Hudson, Bratin Saha, and Adam Welc. Practical weak-atomicity semantics for Java STM. In *SPAA '08: Proc. 20th Annual Symposium on Parallelism in Algorithms and Architectures*, pages 314–325, June 2008. DOI: 10.1145/1378533.1378588 31, 62, 65, 136

[221] Vijay S. Menon, Neal Glew, Brian R. Murphy, Andrew McCreight, Tatiana Shpeisman, Ali-Reza Adl-Tabatabai, and Leaf Petersen. A verifiable SSA program representation for aggressive compiler optimization. In *POPL '06: Proc. 33rd ACM SIGPLAN-SIGACT Symposium on Principles of Programming Languages*, pages 397–408, January 2006. 114

[222] Maged M. Michael. Hazard pointers: safe memory reclamation for lock-free objects. *IEEE Transactions on Parallel and Distributed Systems*, 15(6):491–504, June 2004. DOI: 10.1109/TPDS.2004.8 82

[223] Maged M. Michael and Michael L. Scott. Simple, fast, and practical non-blocking and blocking concurrent queue algorithms. In *PODC '96: Proc. 15th Annual ACM Symposium on Prin-*

ciples of Distributed Computing, pages 267–275, May 1996. DOI: 10.1145/248052.248106 2

[224] Maged M. Michael, Martin T. Vechev, and Vijay A. Saraswat. Idempotent work stealing. In *PPoPP '09: Proc. 14th Symposium on Principles and Practice of Parallel Programming*, pages 45–54, February 2009. DOI: 10.1145/1504176.1504186 2

[225] Miloš Milovanović, Roger Ferrer, Vladimir Gajinov, Osman S. Unsal, Adrián Cristal, Eduard Ayguadé, and Mateo Valero. Multithreaded software transactional memory and OpenMP. In *MEDEA '07: Proc. 2007 Workshop on Memory Performance*, pages 81–88, September 2007. DOI: 10.1145/1327171.1327181 44

[226] Katherine F. Moore and Dan Grossman. High-level small-step operational semantics for transactions. In *POPL '08: Proc. 35th Annual ACM SIGPLAN-SIGACT Symposium on Principles of Programming Languages*, pages 51–62, January 2008. Earlier version presented at *TRANSACT '07*. DOI: 10.1145/1328438.1328448 23, 31, 38, 40, 44, 68, 145

[227] Kevin E. Moore, Jayaram Bobba, Michelle J. Moravan, Mark D. Hill, and David A. Wood. LogTM: Log-based transactional memory. In *HPCA '06: Proc. 12th International Symposium on High-Performance Computer Architecture*, pages 254–265, February 2006. 14, 21, 22, 147, 170, 172

[228] Michelle J. Moravan, Jayaram Bobba, Kevin E. Moore, Luke Yen, Mark D. Hill, Ben Liblit, Michael M. Swift, and David A. Wood. Supporting nested transactional memory in LogTM. In *ASPLOS '06: Proc. 12th International Conference on Architectural Support for Programming Languages and Operating Systems*, pages 359–370, October 2006. DOI: 10.1145/1168857.1168902 81, 201

[229] J. Eliot B. Moss and Antony L. Hosking. Nested transactional memory: Model and architecture sketches. 63(2):186–201, December 2006. DOI: 10.1016/j.scico.2006.05.010 42, 43, 202

[230] Armand Navabi, Xiangyu Zhang, and Suresh Jagannathan. Quasi-static scheduling for safe futures. In *PPoPP '08: Proc. 13th ACM SIGPLAN Symposium on Principles and Practice of Parallel Programming*, pages 23–32, February 2008. DOI: 10.1145/1345206.1345212 98

[231] Yang Ni, Vijay S. Menon, Ali-Reza Adl-Tabatabai, Antony L. Hosking, Richard L. Hudson, J. Eliot B. Moss, Bratin Saha, and Tatiana Shpeisman. Open nesting in software transactional memory. In *PPoPP '07: Proc. 12th ACM SIGPLAN symposium on Principles and Practice of Parallel Programming*, pages 68–78, March 2007. DOI: 10.1145/1229428.1229442 43

[232] Yang Ni, Adam Welc, Ali-Reza Adl-Tabatabai, Moshe Bach, Sion Berkowits, James Cownie, Robert Geva, Sergey Kozhukow, Ravi Narayanaswamy, Jeffrey Olivier, Serguei Preis, Bratin

Saha, Ady Tal, and Xinmin Tian. Design and implementation of transactional constructs for C/C++. In *OOPSLA '08: Proc. 23rd ACM SIGPLAN Conference on Object-Oriented Programming, Systems, Languages, and Applications*, pages 195–212, September 2008. DOI: 10.1145/1449764.1449780 62, 69, 80

[233] Konstantinos Nikas, Nikos Anastopoulos, Georgios Goumas, and Nektarios Koziris. Employing transactional memory and helper threads to speedup Dijkstra's algorithm. In *ICPP '09: Proc. 38th International Conference on Parallel Processing*, September 2009. 91

[234] Cosmin E. Oancea, Alan Mycroft, and Tim Harris. A lightweight in-place implementation for software thread-level speculation. In *SPAA '09: Proc. 21st Symposium on Parallelism in Algorithms and Architectures*, August 2009. DOI: 10.1145/1583991.1584050 xiv

[235] John O'Leary, Bratin Saha, and Mark R. Tuttle. Model checking transactional memory with Spin (brief announcement). In *PODC '08: Proc. 27th ACM symposium on Principles of Distributed Computing*, pages 424–424, August 2008. DOI: 10.1145/1400751.1400816 144

[236] Marek Olszewski, Jeremy Cutler, and J. Gregory Steffan. JudoSTM: a dynamic binary-rewriting approach to software transactional memory. In *PACT '07: Proc. 16th International Conference on Parallel Architecture and Compilation Techniques*, pages 365–375, September 2007. DOI: 10.1109/PACT.2007.42 55, 68, 102, 124, 126

[237] Kunle Olukotun and Lance Hammond. The future of microprocessors. *Queue*, 3(7):26–29, 2005. DOI: 10.1145/1095408.1095418 1

[238] Victor Pankratius, Ali-Reza Adl-Tabatabai, and Frank Otto. Does transactional memory keep its promises? Results from an empirical study. Technical Report 2009-12, IPD, University of Karlsruhe, Germany, September 2009. 94

[239] Salil Pant and Gregory Byrd. Extending concurrency of transactional memory programs by using value prediction. In *CF '09: Proc. 6th ACM conference on Computing frontiers*, pages 11–20, May 2009. DOI: 10.1145/1531743.1531748 56

[240] Salil Pant and Gregory Byrd. Limited early value communication to improve performance of transactional memory. In *ICS '09: Proc. 23rd International Conference on Supercomputing*, pages 421–429, June 2009. DOI: 10.1145/1542275.1542334 56

[241] Cristian Perfumo, Nehir Sönmez, Srdjan Stipic, Osman S. Unsal, Adrián Cristal, Tim Harris, and Mateo Valero. The limits of software transactional memory (STM): dissecting Haskell STM applications on a many-core environment. In *CF '08: Proc. 5th conference on Computing frontiers*, pages 67–78, May 2008. Earlier version presented at *TRANSACT '07*. DOI: 10.1145/1366230.1366241 90, 92

[242] Filip Pizlo, Marek Prochazka, Suresh Jagannathan, and Jan Vitek. Transactional lock-free objects for real-time Java. In *CSJP '04: Proc. ACM PODC Workshop on Concurrency and Synchronization in Java Programs*, pages 54–62, 2004. 53, 63, 84

[243] Donald E. Porter, Owen S. Hofmann, Christopher J. Rossbach, Alexander Benn, and Emmett Witchel. Operating systems transactions. In *SOSP '09: Proc. 22nd ACM SIGOPS Symposium on Operating Systems Principles*, pages 161–176, October 2009. DOI: 10.1145/1629575.1629591 15, 81, 88

[244] Donald E. Porter and Emmett Witchel. Operating systems should provide transactions. In *HotOS '09: Proc. 12th Workshop on Hot Topics in Operating Systems*, May 2009. 88

[245] Seth H. Pugsley, Manu Awasthi, Niti Madan, Naveen Muralimanohar, and Rajeev Balasubramonian. Scalable and reliable communication for hardware transactional memory. In *PACT '08: Proc. 17th International Conference on Parallel Architectures and Compilation Techniques*, pages 144–154, October 2008. DOI: 10.1145/1454115.1454137 182

[246] George Radin. The 801 minicomputer. In *ASPLOS '82: Proc. 1st International Symposium on Architectural Support for Programming Languages and Operating Systems*, pages 39–47, 1982. DOI: 10.1145/800050.801824 194

[247] Ravi Rajwar. *Speculation-Based Techniques for Transactional Lock-Free Execution of Lock-Based Programs*. PhD thesis, October 2002. University of Wisconsin. 96, 150, 160

[248] Ravi Rajwar and James R. Goodman. Speculative lock elision: enabling highly concurrent multithreaded execution. In *MICRO '01: Proc. 34th International Symposium on Microarchitecture*, pages 294–305, December 2001. 93, 96, 149, 150, 159

[249] Ravi Rajwar and James R. Goodman. Transactional lock-free execution of lock-based programs. In *ASPLOS '02: Proc. 10th Symposium on Architectural Support for Programming Languages and Operating Systems*, pages 5–17, October 2002. DOI: 10.1145/605397.605399 159, 161

[250] Ravi Rajwar and James R. Goodman. Transactional execution: toward reliable, high-performance multithreading. *IEEE Micro*, 23(6):117–125, Nov-Dec 2003. DOI: 10.1109/MM.2003.1261395 166

[251] Ravi Rajwar, Maurice Herlihy, and Konrad Lai. Virtualizing transactional memory. In *ISCA '05: Proc. 32nd Annual International Symposium on Computer Architecture*, pages 494–505, June 2005. 191

[252] Hany Ramadan and Emmett Witchel. The Xfork in the road to coordinated sibling transactions. In *TRANSACT '09: 4th Workshop on Transactional Computing*, February 2009. 44

[253] Hany E. Ramadan, Christopher J. Rossbach, Donald E. Porter, Owen S. Hofmann, Aditya Bhandari, and Emmett Witchel. MetaTM/TxLinux: transactional memory for an operating system. In *ISCA '07: Proc. 34th annual international symposium on Computer architecture*, pages 92–103, 2007. A later paper about this work appeared in CACM 51(9), September 2008. DOI: 10.1145/1250662.1250675 93

[254] Hany E. Ramadan, Christopher J. Rossbach, and Emmett Witchel. Dependence-aware transactional memory for increased concurrency. In *MICRO '08: Proc. 2008 41st IEEE/ACM International Symposium on Microarchitecture*, pages 246–257, 2008. DOI: 10.1109/MICRO.2008.4771795 56

[255] Hany E. Ramadan, Indrajit Roy, Maurice Herlihy, and Emmett Witchel. Committing conflicting transactions in an STM. In *PPoPP '09: Proc. 14th ACM SIGPLAN Symposium on Principles and Practice of Parallel Programming*, pages 163–172, February 2009. DOI: 10.1145/1504176.1504201 56

[256] R. Ramakrishnan and J. Gehrke. *Database Management Systems*. McGraw-Hill, 2000. 5

[257] G. Ramalingam. Context-sensitive synchronization-sensitive analysis is undecidable. *TOPLAS: ACM Transactions on Programming Languages and Systems*, 22(2):416–430, 2000. DOI: 10.1145/349214.349241 2

[258] Paruj Ratanaworabhan, Martin Burtscher, Darko Kirovski, Benjamin Zorn, Rahul Nagpal, and Karthik Pattabiraman. Detecting and tolerating asymmetric races. In *PPoPP '09: Proc. 14th ACM SIGPLAN Symposium on Principles and Practice of Parallel Programming*, pages 173–184, February 2009. DOI: 10.1145/1504176.1504202 31

[259] Lukas Renggli and Oscar Nierstrasz. Transactional memory for Smalltalk. In *ICDL '07: Proc. 2007 International Conference on Dynamic Languages*, pages 207–221, August 2007. DOI: 10.1145/1352678.1352692 62

[260] Torval Riegel, Christof Fetzer, Heiko Sturzrehm, and Pascal Felber. From causal to z-linearizable transactional memory (brief announcement). In *PODC '07: Proc. 26th ACM symposium on Principles of distributed computing*, pages 340–341, August 2007. DOI: 10.1145/1281100.1281162 27

[261] Torvald Riegel and Diogo Becker de Brum. Making object-based STM practical in unmanaged environments. In *TRANSACT '08: 3rd Workshop on Transactional Computing*, February 2008. 105

[262] Torvald Riegel, Pascal Felber, and Christof Fetzer. A lazy snapshot algorithm with eager validation. In *DISC '06: Proc. 20th International Symposium on Distributed Computing*, volume 4167 of *Lecture Notes in Computer Science*, pages 284–298, September 2006. DOI: 10.1007/11864219_20 15, 123

[263] Torvald Riegel, Christof Fetzer, and Pascal Felber. Snapshot isolation for software transactional memory. In *TRANSACT '06: 1st Workshop on Languages, Compilers, and Hardware Support for Transactional Computing*, June 2006. 27, 123

[264] Torvald Riegel, Christof Fetzer, and Pascal Felber. Time-based transactional memory with scalable time bases. In *SPAA '07: Proc. 19th ACM Symposium on Parallelism in Algorithms and Architectures*, pages 221–228, June 2007. DOI: 10.1145/1248377.1248415 121

[265] Torvald Riegel, Christof Fetzer, and Pascal Felber. Automatic data partitioning in software transactional memories. In *SPAA '08: Proc. 20th Symposium on Parallelism in Algorithms and Architectures*, pages 152–159, June 2008. DOI: 10.1145/1378533.1378562 105

[266] Michael F. Ringenburg and Dan Grossman. AtomCaml: First-class atomicity via rollback. In *Proc. 10th ACM SIGPLAN International Conference on Functional Programming*, September 2005. DOI: 10.1145/1086365.1086378 62, 73, 75, 86

[267] Paolo Romano, Nuno Carvalho, and Luís Rodrigues. Towards distributed software transactional memory systems. In *LADIS '08: Proc. 2nd Workshop on Large-Scale Distributed Systems and Middleware*, September 2008. DOI: 10.1145/1529974.1529980 143

[268] Daniel J. Rosenkrantz, Richard E. Stearns, and Philip M. Lewis, II. System level concurrency control for distributed database systems. *TODS: ACM Transactions on Database Systems*, 3(2):178–198, 1978. DOI: 10.1145/320251.320260 161

[269] Christopher Rossbach, Owen Hofmann, and Emmett Witchel. Is transactional memory programming actually easier? In *PPoPP '10: Proc. 15th ACM SIGPLAN Symposium on Principles and Practice of Parallel Programming*, pages 47–56, January 2010. Earlier version presented at *TRANSACT '09*. DOI: 10.1145/1693453.1693462 93

[270] Christopher J. Rossbach, Owen S. Hofmann, Donald E. Porter, Hany E. Ramadan, Aditya Bhandari, and Emmett Witchel. TxLinux: using and managing hardware transactional memory in an operating system. In *SOSP '07: Proc. 21st ACM SIGOPS Symposium on Operating Systems Principles*, pages 87–102, October 2007. A later paper about this work appeared in *CACM* 51(9), September 2008. DOI: 10.1145/1294261.1294271 93

[271] Christopher J. Rossbach, Hany E. Ramadan, Owen S. Hofmann, Donald E. Porter, Aditya Bhandari, and Emmett Witchel. TxLinux and MetaTM: transactional memory and the operating system. *Communications of the ACM*, 51(9):83–91, September 2008. Earlier versions of this work appeared at *ISCA '07* and *SOSP '07*. DOI: 10.1145/1378727.1378747 14, 93

[272] Amitabha Roy, Steven Hand, and Tim Harris. A runtime system for software lock elision. In *EuroSys '09: Proc. 4th ACM European Conference on Computer Systems*, pages 261–274, April 2009. DOI: 10.1145/1519065.1519094 96

[273] Bratin Saha, Ali-Reza Adl-Tabatabai, Anwar Ghuloum, Mohan Rajagopalan, Richard L. Hudson, Leaf Petersen, Vijay Menon, Brian Murphy, Tatiana Shpeisman, Eric Sprangle, Anwar Rohillah, Doug Carmean, and Jesse Fang. Enabling scalability and performance in a large scale CMP environment. In *EuroSys '07: Proc. 2nd ACM SIGOPS/EuroSys European Conference on Computer Systems 2007*, pages 73–86, March 2007. DOI: 10.1145/1272996.1273006 109

[274] Bratin Saha, Ali-Reza Adl-Tabatabai, Richard L. Hudson, Chi Cao Minh, and Benjamin Hertzberg. McRT-STM: a high performance software transactional memory system for a multi-core runtime. In *PPoPP '06: Proc. 11th ACM SIGPLAN Symposium on Principles and Practice of Parallel Programming*, pages 187–197, March 2006. DOI: 10.1145/1122971.1123001 47, 102, 108, 109, 115

[275] Bratin Saha, Ali-Reza Adl-Tabatabai, and Quinn Jacobson. Architectural support for software transactional memory. In *MICRO '06: Proc. 39th Annual IEEE/ACM International Symposium on Microarchitecture*, pages 185–196, 2006. DOI: 10.1109/MICRO.2006.9 197

[276] Daniel Sanchez, Luke Yen, Mark D. Hill, and Karthikeyan Sankaralingam. Implementing signatures for transactional memory. In *MICRO '07: Proc. 40th Annual IEEE/ACM International Symposium on Microarchitecture*, pages 123–133, 2007. DOI: 10.1109/MICRO.2007.20 174, 175

[277] Sutirtha Sanyal, Adrián Cristal, Osman S. Unsal, Mateo Valero, and Sourav Roy. Dynamically filtering thread-local variables in lazy-lazy hardware transactional memory. In *HPCC '09: Proc. 11th Conference on High Performance Computing and Communications*, June 2009. 178

[278] William N. Scherer III and Michael L. Scott. Contention management in dynamic software transactional memory. In *CSJP '04: Proc. ACM PODC Workshop on Concurrency and Synchronization in Java Programs*, July 2004. In conjunction with PODC'04. Please also download errata from http://www.cs.rochester.edu/u/scott/papers/ 2004_CSJP_contention_mgmt_errata.pdf. 29, 51, 131

[279] William N. Scherer III and Michael L. Scott. Nonblocking concurrent objects with condition synchronization. In *DISC '04: Proc. 18th International Symposium on Distributed Computing*, October 2004. 75

[280] William N. Scherer III and Michael L. Scott. Advanced contention management for dynamic software transactional memory. In *PODC '05: Proc. 24th ACM Symposium on Principles of Distributed Computing*, pages 240–248, July 2005. DOI: 10.1145/1073814.1073861 51, 52

[281] Florian T. Schneider, Vijay Menon, Tatiana Shpeisman, and Ali-Reza Adl-Tabatabai. Dynamic optimization for efficient strong atomicity. In *OOPSLA '08: Proc. 23rd ACM SIGPLAN Conference on Object-Oriented Programming, Systems, Languages, and Applications*, pages 181–194, September 2008. DOI: 10.1145/1449764.1449779 35, 71

[282] Michael L. Scott. Sequential specification of transactional memory semantics. In *TRANS-ACT '06: 1st Workshop on Languages, Compilers, and Hardware Support for Transactional Computing*, June 2006. 23

[283] Michael L. Scott, Michael F. Spear, Luke Dalessandro, and Virendra J. Marathe. Delaunay triangulation with transactions and barriers. In *IISWC '07: Proc. 2007 IEEE International Symposium on Workload Characterization*, September 2007. Benchmarks track. DOI: 10.1109/IISWC.2007.4362186 91

[284] Michael L. Scott, Michael F. Spear, Luke Dalessandro, and Virendra J. Marathe. Transactions and privatization in Delaunay triangulation (brief announcement). In *PODC '07: Proc. 26th PODC ACM Symposium on Principles of Distributed Computing*, August 2007. DOI: 10.1145/1281100.1281160 40

[285] Nir Shavit and Dan Touitou. Software transactional memory. In *PODC '95: Proc. 14th ACM Symposium on Principles of Distributed Computing*, pages 204–213, August 1995. DOI: 10.1145/224964.224987 101, 128

[286] Avraham Shinnar, David Tarditi, Mark Plesko, and Bjarne Steensgaard. Integrating support for undo with exception handling. Technical Report MSR-TR-2004-140, Microsoft Research, December 2004. 80, 83

[287] Tatiana Shpeisman, Ali-Reza Adl-Tabatabai, Robert Geva, Yang Ni, and Adam Welc. Towards transactional memory semantics for C++. In *SPAA '09: Proc. 21st Symposium on Parallelism in Algorithms and Architectures*, pages 49–58, August 2009. DOI: 10.1145/1583991.1584012 62, 80

[288] Tatiana Shpeisman, Vijay Menon, Ali-Reza Adl-Tabatabai, Steve Balensiefer, Dan Grossman, Richard Hudson, Katherine F. Moore, and Bratin Saha. Enforcing isolation and ordering in STM. In *PLDI '07: Proc. 2007 ACM SIGPLAN Conference on Programming Language Design and Implementation*, pages 78–88, June 2007. DOI: 10.1145/1250734.1250744 31, 35, 40, 71

[289] Arrvindh Shriram, Virendra J. Marathe, Sandhya Dwarkadas, Michael L. Scott, David Eisenstat, Christopher Heriot, William N. Scherer III, and Michael F. Spear. Hardware acceleration of software transactional memory. Technical Report TR 887, Computer Science Department, University of Rochester, December 2005. Revised, March 2006; condensed version presented at TRANSACT '06. 182, 198

[290] Arrvindh Shriraman and Sandhya Dwarkadas. Refereeing conflicts in hardware transactional memory. In *ICS '09: Proc. 23rd International Conference on Supercomputing*, pages 136–146, June 2009. Also available as TR 939, Department of Computer Science, University of Rochester, September 2008. DOI: 10.1145/1542275.1542299 50, 52, 147, 152, 170, 182, 198

[291] Arrvindh Shriraman, Sandhya Dwarkadas, and Michael L. Scott. Flexible decoupled transactional memory support. *Journal of Parallel and Distributed Computing: Special Issue on Transactional Memory*, June 2010. Earlier version published in *ISCA '08*. 50, 52, 147, 170, 182, 183, 198

[292] Arrvindh Shriraman, Michael F. Spear, Hemayet Hossain, Virendra Marathe, Sandhya Dwarkadas, and Michael L. Scott. An integrated hardware-software approach to flexible transactional memory. In *ISCA '07: Proc. 34rd Annual International Symposium on Computer Architecture*, pages 104–115, June 2007. DOI: 10.1145/1250662.1250676 197

[293] Ed Sikha, Rick Simpson, Cathy May, and Hank Warren, editors. *The PowerPC architecture: a specification for a new family of RISC processors*. Morgan Kaufmann Publishers Inc., San Francisco, CA, USA, 1994. 155

[294] Satnam Singh. Higher order combinators for join patterns using STM. In *TRANSACT '06: 1st Workshop on Languages, Compilers, and Hardware Support for Transactional Computing*, June 2006. 92

[295] Travis Skare and Christos Kozyrakis. Early release: friend or foe? In *Proc. Workshop on Transactional Workloads*, June 2006. 56

[296] J. E. Smith and G. Sohi. The microarchitecture of superscalar processors. *Proc. IEEE*, 48:1609–1624, December 1995. DOI: 10.1109/5.476078 152, 159

[297] Nehir Sonmez, Tim Harris, Adrián Cristal, Osman S. Unsal, and Mateo Valero. Taking the heat off transactions: dynamic selection of pessimistic concurrency control. In *IPDPS '09: Proc. 23rd International Parallel and Distributed Processing Symposium*, May 2009. 54, 109

[298] Nehir Sonmez, Cristian Perfumo, Srdan Stipic, Osman Unsal, Adrian Cristal, and Mateo Valero. UnreadTVar: extending Haskell software transactional memory for performance. In *Symposium on Trends in Functional Programming*, April 2007. 56

[299] Michael F. Spear, Luke Dalessandro, Virendra Marathe, and Michael L. Scott. Ordering-based semantics for software transactional memory. In *OPODIS '08: Proc. 12th International Conference on Principles of Distributed Systems*, pages 275–294, December 2008. Springer-Verlag Lecture Notes in Computer Science volume 5401. DOI: 10.1007/978-3-540-92221-6_19 38, 136

[300] Michael F. Spear, Luke Dalessandro, Virendra J. Marathe, and Michael L. Scott. A comprehensive strategy for contention management in software transactional memory. In *PPoPP '09: Proc. 14th ACM SIGPLAN Symposium on Principles and Practice of Parallel Programming*, pages 141–150, February 2009. DOI: 10.1145/1504176.1504199 21, 50, 52, 53, 107, 108, 121, 184

[301] Michael F. Spear, Virendra J. Marathe, Luke Dalessandro, and Michael L. Scott.   Privatization techniques for software transactional memory (brief announcement).   In *PODC '07: Proc. 26th PODC ACM Symposium on Principles of Distributed Computing*, August 2007.   Extended version available as TR-915, Computer Science Department, University of Rochester, Feb. 2007, http://www.cs.rochester.edu/u/scott/papers/2007_TR915.pdf. DOI: 10.1145/1281100.1281161 137

[302] Michael F. Spear, Virendra J. Marathe, William N. Scherer III, and Michael L. Scott. Conflict detection and validation strategies for software transactional memory. In *DISC '06: Proc. 20th International Symposium on Distributed Computing*, September 2006. 115, 123, 131, 132

[303] Michael F. Spear, Maged M. Michael, and Michael L. Scott. Inevitability mechanisms for software transactional memory. In *TRANSACT '08: 3rd Workshop on Transactional Computing*, February 2008. 21, 81, 87

[304] Michael F. Spear, Maged M. Michael, Michael L. Scott, and Peng Wu. Reducing memory ordering overheads in software transactional memory.   In *CGO '09: Proc. 2009 International Symposium on Code Generation and Optimization*, pages 13–24, March 2009. DOI: 10.1109/CGO.2009.30 103, 120

[305] Michael F. Spear, Maged M. Michael, and Christoph von Praun. RingSTM: scalable transactions with a single atomic instruction. In *SPAA '08: Proc. 20th Annual Symposium on Parallelism in Algorithms and Architectures*, pages 275–284, June 2008. DOI: 10.1145/1378533.1378583 15, 102, 124

[306] Michael F. Spear, Arrvindh Shriraman, Luke Dalessandro, Sandhya Dwarkadas, and Michael L. Scott. Nonblocking transactions without indirection using alert-on-update. In *SPAA '07: Proc. 19th ACM Symposium on Parallel Algorithms and Architectures*, pages 210–220, June 2007. DOI: 10.1145/1248377.1248414 134

[307] Michael F. Spear, Arrvindh Shriraman, Luke Dalessandro, and Michael L. Scott. Transactional mutex locks. In *TRANSACT '09: 4th Workshop on Transactional Computing*, February 2009. 126

[308] Michael F. Spear, Michael Silverman, Luke Dalessandro, Maged M. Michael, and Michael L. Scott.   Implementing and exploiting inevitability in software transactional memory.   In *ICPP '08: Proc. 37th International Conference on Parallel Processing*, September 2008. DOI: 10.1109/ICPP.2008.55 21, 87, 91, 141, 142

[309] Janice M. Stone, Harold S. Stone, Phil Heidelberger, and John Turek. Multiple reservations and the Oklahoma update. *IEEE Parallel & Distributed Technology*, 1(4):58–71, November 1993. DOI: 10.1109/88.260295 6, 149, 150, 158

[310] P. Sweazey and A. J. Smith. A class of compatible cache consistency protocols and their support by the IEEE Futurebus. In *ISCA '86: Proc. 13th annual international symposium on Computer architecture*, pages 414–423, 1986. 151

[311] Michael Swift, Haris Volos, Neelam Goyal, Luke Yen, Mark Hill, and David Wood. OS support for virtualizing hardware transactional memory. In *TRANSACT '08: 3rd Workshop on Transactional Computing*, February 2008. 187

[312] Fuad Tabba, Andrew W. Hay, and James R. Goodman. Transactional value prediction. In *TRANSACT '09: 4th Workshop on Transactional Computing*, February 2009. 55

[313] Fuad Tabba, Mark Moir, James R. Goodman, Andrew Hay, and Cong Wang. NZTM: Nonblocking zero-indirection transactional memory. In *SPAA '09: Proc. 21st Symposium on Parallelism in Algorithms and Architectures*, August 2009. 47, 102, 134, 135

[314] Fuad Tabba, Cong Wang, James R. Goodman, and Mark Moir. NZTM: Nonblocking, zero-indirection transactional memory. In *TRANSACT '07: 2nd Workshop on Transactional Computing*, August 2007. 47, 102, 134, 135

[315] Serdar Tasiran. A compositional method for verifying software transactional memory implementations. Technical Report MSR-TR-2008-56, Microsoft Research, April 2008. 145

[316] Rubén Titos, Manuel E. Acacio, and José M. García. Directory-based conflict detection in hardware transactional memory. In *HiPC '08: Proc. 15th International Conference on High Performance Computing*, December 2008. Springer-Verlag Lecture Notes in Computer Science volume 5374. DOI: 10.1007/978-3-540-89894-8_47 164

[317] Rubén Titos, Manuel E. Acacio, and Jose M. Garcia. Speculation-based conflict resolution in hardware transactional memory. In *IPDPS '09: Proc. 23rd International Parallel and Distributed Processing Symposium*, May 2009. 56

[318] Sasa Tomic, Cristian Perfumo, Chinmay Kulkarni, Adria Armejach, Adrián Cristal, Osman Unsal, Tim Harris, and Mateo Valero. EazyHTM: eager-lazy hardware transactional memory. In *MICRO '09: Proc. 2009 42nd IEEE/ACM International Symposium on Microarchitecture*, December 2009. DOI: 10.1145/1669112.1669132 50, 184

[319] Takayuki Usui, Yannis Smaragdakis, and Reimer Behrends. Adaptive locks: Combining transactions and locks for efficient concurrency. In *PACT '09: Proc. 18th International Conference on Parallel Architectures and Compilation Techniques*, September 2009. 98

[320] Jan Vitek, Suresh Jagannathan, Adam Welc, and Antony L. Hosking. A semantic framework for designer transactions. In *ESOP '04: Proc. European Symposium on Programming*, volume 2986 of *Lecture Notes in Computer Science*, pages 249–263, 2004. DOI: 10.1007/b96702 44

[321] Haris Volos, Neelam Goyal, and Michael Swift. Pathological interaction of locks with transactional memory. In *TRANSACT '08: 3rd Workshop on Transactional Computing*, February 2008. 84, 203

[322] Haris Volos, Andres Jaan Tack, Neelam Goyal, Michael M. Swift, and Adam Welc. xCalls: safe I/O in memory transactions. In *EuroSys '09: Proc. 4th ACM European Conference on Computer Systems*, pages 247–260, April 2009. DOI: 10.1145/1519065.1519093 81, 88

[323] Haris Volos, Adam Welc, Ali-Reza Adl-Tabatabai, Tatiana Shpeisman, Xinmin Tian, and Ravi Narayanaswamy. NePaLTM: Design and implementation of nested parallelism for transactional memory systems. In *ECOOP '09: Proc. 23rd European Conference on Object-Oriented Programming*, June 2009. Springer-Verlag Lecture Notes in Computer Science volume 5653. DOI: 10.1007/978-3-642-03013-0_7 44

[324] Christoph von Praun, Luis Ceze, and Calin Caşcaval. Implicit parallelism with ordered transactions. In *PPoPP '07: Proc. 12th ACM SIGPLAN symposium on Principles and Practice of Parallel Programming*, pages 79–89, March 2007. DOI: 10.1145/1229428.1229443 98, 179

[325] M. M. Waliullah and Per Stenström. Starvation-free transactional memory system protocols. In *Proc. 13th Euro-Par Conference: European Conference on Parallel and Distributed Computing*, pages 280–291, August 2007. 182

[326] M. M. Waliullah and Per Stenström. Intermediate checkpointing with conflicting access prediction in transactional memory systems. In *IPDPS '08: Proc. 22nd International Parallel and Distributed Processing Symposium*, April 2008. DOI: 10.1109/IPDPS.2008.4536249 57

[327] Cheng Wang, Wei-Yu Chen, Youfeng Wu, Bratin Saha, and Ali-Reza Adl-Tabatabai. Code generation and optimization for transactional memory constructs in an unmanaged language. In *CGO '07: Proc. International Symposium on Code Generation and Optimization*, pages 34–48, March 2007. DOI: 10.1109/CGO.2007.4 62, 68, 105, 120, 121, 137

[328] Cheng Wang, Victor Ying, and Youfeng Wu. Supporting legacy binary code in a software transaction compiler with dynamic binary translation and optimization. In *CC '08: Proc. International Conference on Compiler Construction*, pages 291–306, March 2008. DOI: 10.1007/978-3-540-78791-4_20 68

[329] William E. Weihl. Data-dependent concurrency control and recovery (extended abstract). In *PODC '83: Proc. second Annual ACM symposium on Principles of distributed computing*, pages 63–75, 1983. DOI: 10.1145/800221.806710 58

[330] Adam Welc, Antony L. Hosking, and Suresh Jagannathan. Transparently reconciling transactions with locking for Java synchronization. In *ECOOP '06: Proc. European Conference on Object-Oriented Programming*, pages 148–173, July 2006. DOI: 10.1007/11785477 96, 97

[331] Adam Welc, Suresh Jagannathan, and Antony Hosking. Safe futures for Java. In *OOPSLA '05: Proc. 20th ACM SIGPLAN Conference on Object-Oriented Programming, Systems, Languages, and Applications*, pages 439–453, October 2005. DOI: 10.1145/1094811.1094845 98

[332] Adam Welc, Suresh Jagannathan, and Antony L. Hosking. Transactional monitors for concurrent objects. In *ECOOP '04: Proc. European Conference on Object-Oriented Programming*, volume 3086 of *Lecture Notes in Computer Science*, pages 519–542, 2004. DOI: 10.1007/b98195 50

[333] Adam Welc, Bratin Saha, and Ali-Reza Adl-Tabatabai. Irrevocable transactions and their applications. In *SPAA '08: Proc. 20th Annual Symposium on Parallelism in Algorithms and Architectures*, pages 285–296, June 2008. DOI: 10.1145/1378533.1378584 21, 81, 87, 141, 142

[334] I. W. Williams and M. I. Wolczko. An object-based memory architecture. In *Proc. 4th International Workshop on Persistent Object Systems*, pages 114–130. Morgan Kaufmann, 1990. 150

[335] Greg Wright, Matthew L. Seidl, and Mario Wolczko. An object-aware memory architecture. *Science of Computer Programming*, 62(2):145–163, 2006. DOI: 10.1016/j.scico.2006.02.007 150

[336] Peng Wu, Maged M. Michael, Christoph von Praun, Takuya Nakaike, Rajesh Bordawekar, Harold W. Cain, Calin Cascaval, Siddhartha Chatterjee, Stefanie Chiras, Rui Hou, Mark F. Mergen, Xiaowei Shen, Michael F. Spear, Huayong Wang, and Kun Wang. Compiler and runtime techniques for software transactional memory optimization. *Concurrency and Computation: Practice and Experience*, 21(1):7–23, 2009. DOI: 10.1002/cpe.1336 62, 114

[337] Luke Yen, Jayaram Bobba, Michael M. Marty, Kevin E. Moore, Haris Volos, Mark D. Hill, Michael M. Swift, and David A. Wood. LogTM-SE: Decoupling hardware transactional memory from caches. In *HPCA '07: Proc. 13th International Symposium on High-Performance Computer Architecture*, February 2007. DOI: 10.1109/HPCA.2007.346204 14, 177, 178, 185, 186

[338] Luke Yen, Stark C. Draper, and Mark D. Hill. Notary: Hardware techniques to enhance signatures. In *MICRO '08: Proc. 2008 41st IEEE/ACM International Symposium on Microarchitecture*, pages 234–245, 2008. DOI: 10.1109/MICRO.2008.4771794 178

[339] V. Ying, C. Wang, Y. Wu, and X. Jiang. Dynamic binary translation and optimization of legacy library code in a STM compilation environment. In *Proc. Workshop on Binary Instrumentation and Applications*, October 2006. 68

[340] Richard M. Yoo and Hsien-Hsin S. Lee. Adaptive transaction scheduling for transactional memory systems. In *SPAA '08: Proc. 20th Annual Symposium on Parallelism in Algorithms and Architectures*, pages 169–178, June 2008. DOI: 10.1145/1378533.1378564 53

[341] Richard M. Yoo, Yang Ni, Adam Welc, Bratin Saha, Ali-Reza Adl-Tabatabai, and Hsien-Hsin S. Lee. Kicking the tires of software transactional memory: why the going gets tough. In *SPAA '08: Proc. 20th Annual Symposium on Parallelism in Algorithms and Architectures*, pages 265–274, June 2008. DOI: 10.1145/1378533.1378582 139

[342] Rui Zhang, Zoran Budimlić, and William N. Scherer III. Commit phase in timestamp-based STM. In *SPAA '08: Proc. 20th Annual Symposium on Parallelism in Algorithms and Architectures*, pages 326–335, June 2008. DOI: 10.1145/1378533.1378589 121, 122

[343] Lukasz Ziarek and Suresh Jagannathan. Memoizing multi-threaded transactions. In *DAMP '08: Proc. Workshop on Declarative Aspects of Multicore Programming*, 2008. 92

[344] Lukasz Ziarek, K. C. Sivaramakrishnan, and Suresh Jagannathan. Partial memoization of concurrency and communication. In *ICFP '09: Proc. 14th ACM SIGPLAN International Conference on Functional Programming*, pages 161–172, August 2009. DOI: 10.1145/1596550.1596575 92

[345] Lukasz Ziarek, Adam Welc, Ali-Reza Adl-Tabatabai, Vijay Menon, Tatiana Shpeisman, and Suresh Jagannathan. A uniform transactional execution environment for Java. In *ECOOP '08: Proc. 22nd European Conference on Object-Oriented Programming*, pages 129–154, July 2008. Springer-Verlag Lecture Notes in Computer Science volume 5142. DOI: 10.1007/978-3-540-70592-5_7 62, 96, 97

[346] Craig Zilles and Lee Baugh. Extending hardware transactional memory to support non-busy waiting and nontransactional actions. In *TRANSACT '06: 1st Workshop on Languages, Compilers, and Hardware Support for Transactional Computing*, June 2006. 43, 81, 88, 202

[347] Craig Zilles and David Flint. Challenges to providing performance isolation in transactional memories. In *WDDD '05: Proc. 4th Workshop on Duplicating, Deconstructing, and Debunking*, pages 48–55, June 2005. 51, 164

[348] Craig Zilles and Ravi Rajwar. Implications of false conflict rate trends for robust software transactional memory. In *IISWC '07: Proc. 2007 IEEE INTL Symposium on Workload Characterization*, September 2007. DOI: 10.1109/IISWC.2007.4362177 105, 150

[349] Ferad Zyulkyarov, Adrián Cristal, Sanja Cvijic, Eduard Ayguadé, Mateo Valero, Osman S. Unsal, and Tim Harris. WormBench: a configurable workload for evaluating transactional memory systems. In *MEDEA '08: Proc. 9th Workshop on Memory Performance*, pages 61–68, October 2008. DOI: 10.1145/1509084.1509093 91

[350] Ferad Zyulkyarov, Vladimir Gajinov, Osman S. Unsal, Adrián Cristal, Eduard Ayguadé, Tim Harris, and Mateo Valero. Atomic Quake: using transactional memory in an interactive multiplayer game server. In *PPoPP '09: Proc. 14th ACM SIGPLAN Symposium on Principles and Practice of Parallel Programming*, pages 25–34, February 2009. DOI: 10.1145/1504176.1504183 12, 92

[351] Ferad Zyulkyarov, Tim Harris, Osman S. Unsal, Adrián Cristal, and Mateo Valero. Debugging programs that use atomic blocks and transactional memory. In *PPoPP '10: Proc. 15th ACM SIGPLAN Symposium on Principles and Practice of Parallel Programming*, pages 57–66, January 2010. DOI: 10.1145/1693453.1693463 89, 90

Authors' Biographies

TIM HARRIS

Tim Harris is a Senior Researcher at MSR Cambridge where he works on abstractions for using multi-core computers. He has worked on concurrent algorithms and transactional memory for over ten years, most recently, focusing on the implementation of STM for multi-core computers and the design of programming language features based on it. Harris is currently working on the Barrelfish operating system and on architecture support for programming language runtime systems. Harris has a BA and PhD in computer science from Cambridge University Computer Laboratory. He was on the faculty at the Computer Laboratory from 2000-2004 where he led the department's research on concurrent data structures and contributed to the Xen virtual machine monitor project. He joined Microsoft Research in 2004.

JAMES LARUS

James Larus is Director of Research and Strategy of the Extreme Computing Group in Microsoft Research. Larus has been an active contributor to the programming languages, compiler, and computer architecture communities. He has published many papers and served on numerous program committees and NSF and NRC panels. Larus became an ACM Fellow in 2006. Larus joined Microsoft Research as a Senior Researcher in 1998 to start and, for five years, led the Software Productivity Tools (SPT) group, which developed and applied a variety of innovative techniques in static program analysis and constructed tools that found defects (bugs) in software. This group's research has both had considerable impact on the research community, as well as being shipped in Microsoft products such as the Static Driver Verifier and FX/Cop, and other, widely-used internal software development tools. Larus then became the Research Area Manager for programming languages and tools and started the Singularity research project, which demonstrated that modern programming languages and software engineering techniques could fundamentally improve software architectures. Before joining Microsoft, Larus was an Assistant and Associate Professor of Computer Science at the University of Wisconsin-Madison, where he published approximately 60 research papers and co-led the Wisconsin Wind Tunnel (WWT) research project with Professors Mark Hill and David Wood. WWT was a DARPA and NSF-funded project investigated new approaches to simulating, building, and programming parallel shared-memory computers. Larus's research spanned a number of areas: including new and efficient techniques for measuring and recording executing programs' behavior, tools for analyzing and manipulating compiled and linked programs, programming languages for parallel computing, tools for verifying program correctness, and techniques for compiler analysis and optimization. Larus received his MS and PhD in Computer Science from the University of California, Berkeley in 1989, and an AB in Applied Mathematics from Harvard in 1980. At Berkeley, Larus developed one of the first systems to analyze Lisp programs and determine how to best execute them on a parallel computer.

RAVI RAJWAR

Ravi Rajwar is an architect in Microprocessor and Graphics Development as part of the Intel Architecture Group at Intel Corporation. His research interests include theoretical and practical aspects of computer architecture. In the past, he has investigated resource-efficient microprocessors and architectural support for improving programmability of parallel software. Rajwar received a PhD from the University of Wisconsin-Madison in 2002, a MS from the University of Wisconsin-Madison in 1998, and a BE from the University of Roorkee, India, in 1994, all in Computer Science.

Printed in the United States
by Baker & Taylor Publisher Services